Mapping World Communication

Mapping World Communication

War, Progress, Culture

Armand Mattelart

Translated by Susan Emanuel and James A. Cohen

University of Minnesota Press
Minneapolis
London

The University of Minnesota Press gratefully acknowledges assistance from the French Ministry of Culture toward publication of this work.

Originally published as *La Communication-monde. Histoire des idées et des stratégies,* copyright 1991 by Editions La Découverte, Paris.

Library of Congress Cataloging-in-Publication Data
Mattelart, Armand.
   [Communication-monde. English]
   Mapping world communication : war, progress, culture / Armand
Mattelart ; translation by Susan Emanuel and James A. Cohen.
     p. cm.
   Includes bibliographical references and index.
   ISBN 0-8166-2261-2. — ISBN 0-8166-2262-0 (pbk.)
    1. Communication, International.   2. Communication—History.
I. Title.
P96.I5M32913   1994
302.2—dc20   93-32250

Published by the University of Minnesota Press
2037 University Avenue Southeast, Minneapolis, MN 55455-3092
Printed in the United States of America on acid-free paper

Published by the University Minnesota Press
111 Third Avenue South, Suite 290
Minneapolis, MN 55401-2520

http://www.upress.umn.edu

Third printing, 2002

Contents

Preface to the English Edition

Modern nationalism appeared in continental Europe at the end of the eighteenth century, in the wake of the French Revolution. This was also the era of a new, modern type of imperialist aggression. In 1798, the expeditionary army of General Bonaparte, the future Emperor Napoleon I, landed in Alexandria on the political pretext of striking at England through India. He was accompanied by an auxiliary corps composed of more than 150 artists, engineers, and scholars of all disciplines. Their mission was "progress and the propagation of the Enlightenment in Egypt." For over four years, under military escort, these shock troops of specialists, gathered at the Institut d'Egypte, drew up an inventory of riches—past, present, and potential—of the Land of the Pharaohs, endeavoring to make people "forget, through the benefits of peace, the miseries of conquest."

Communication was at the vanguard of this effort. Participating in the adventure were 23 printers; an "Orientalist" directed the Oriental Printing Office; its two presses, one of which was equipped with Arabic characters borrowed from the Vatican, printed proclamations and bulletins in Arabic, Greek, and Turkish, and two periodicals in French. Optical telegraph stations—which had just been invented and put into use in France—were built. A commission was created to study the feasibility of a project for a great canal from Suez to the Mediterranean.

Cut to the end of the nineteenth century, when communication was consecrated as the "agent of civilization." Its universality was that of the Victorian Empire. From railway networks, the electric telegraph, and underwater cable as well as the new Suez interoceanic route and steam navigation was stitched together an image of the world as a vast "organism," all of whose parts were in solidarity. Networks covering the globe became the symbol of an interdependent world where domestic economies had given way to a new international division of labor.

"All men are brothers" proclaimed the great universal expositions. But outside the microcosms of these so-called peaceful assemblies of progress under the aegis of Western civilization, the biological solidarity among peoples was ceaselessly belied by the sound and fury of wars. It is true that the end of the century saw the emergence of new forms of international association, both among civil societies and among states. But the centennial year of the Egyptian expedition closed with the landing of U.S. Marines in Cuba, under the pretext of helping the natives free themselves from the moribund Spanish empire. For the first time in history, public opinion, stirred up by a sensational press, became the alibi for an imperial intervention.

Cut now to the end of the twentieth century: communication becomes the paradigm of the new global society in an economy based on nonmaterial flows. The space of production and trade spans the world. Two major events underscore the current transformation of the world order: the crumbling of the "socialist bloc"—the end of a certain conception of internationalism—and the Gulf War, symptom of a new world hegemony. The planet begins to worry about the destructive effects of ideologies of progress on the biosphere and utopias are no longer in sight.

The climax of the fin de siècle spectacle: the Universal Exposition of 1992 in Seville, a true shop-window of global communications. In this technological Disneyland, built on the banks of the Guadalquivir River, where Christopher Columbus's three caravels lifted anchor on their voyage to the West Indies, the segregation of the world appeared in all its aggressiveness. So-called postmodern nations and multinational electronics firms—the firms represented alongside the states—astounded visitors with the latest performances of their high technologies of image and information. The less-well-off countries were housed in five collective pavilions, grouped geographically and built at the expense of Spain; they exhibited a few traces of the rich past of their traditions and cultures, but left an impression of a future disconnected from technological innovations.

This book retraces the historical stages of this process of the internationalization of a mode of communication that has progressively become a way of organizing the world. It examines the symbolic and material balances of forces that express themselves in circuits of exchange and flows of people, goods, and services. Such an examination requires us to challenge a concept of communication tied exclusively to the mass media.

Who can deny that "the global" is a fact and that the process of expansion of new technologies of information and communication is only understandable at a world level? But "the global" can also adopt the guise of reductive ideologies. If I have chosen to place this book under the sign of the concept of "world communication," it is intended as a break with reductive mindsets that ignore and sweep under the rug all the fractures and asperities of the world. Thanks to its historical perspective and its kinship with Fernand Braudel's concept of "world economy," this concept seems to me the best able to integrate within it all the necessary levels of analysis. The world is a contradictory system made up all at once of interdependencies and interconnections, of schisms, fragmentations, and exclusions. If the new global configurations are marked by the logics of globalization and homogeneity, they also harbor contrary and interfering logics. To better understand the tension and the new complexity of a world crisscrossed by struggles for hegemony in the giving of meaning to the world, it is fitting, following the philosopher Maurice Merleau-Ponty, to adopt the notion of "baroque systems." This notion reintroduces movement where we had been led to believe in the ineluctability of the reproduction of macrostructures and in the corollary powerlessness of subordinate cultures to express themselves and to avoid being reduced to non-sense and nonbeing.

My framework for analyzing this world communication involves multiple entries. It is, first of all, the intersection of two lines of investigation, a thematic approach and a chronological approach, which inevitably overlap. This is the case, for example, in the first two chapters, which open with the nineteenth century. As we try to show, the theme of war and its relation with communications cannot be understood unless it is tied to the notion of progress and civilization—the major theme of Part II. The "miseries of conquest" go together with the "benefits of peace"!

This study also involves a choice of examples and details as revelatory of the whole. Thus one will not find here exhaustive lists of events or a systematic account of the historical path taken by each component of communications systems. We will, however, trace the sinuous itinerary of each technical object (the telegraph, the underwater cable, the telephone, the railway, radio, television, satellites, etc.) taken as an example, in order to show how it has participated in the formation of world communications.

Finally, this is a history of theories and the ways in which they have been mobilized. Intelligence, propaganda, psychological warfare, disinformation, modernization, the cultural industry (or industries), the information society, interdependence, cultural imperialism, and globalization: this book invites the reader to investigate the sociohistorical context in which each of these concepts appeared and the precise function of each at given moments.

This epistemological examination of the various historical sedimentations of concepts is all the more important in that one of the major problems affecting contemporary research on communication is amnesia, the absence of a "collective memory," the forgetting of the social and political stakes at issue in these theoretical constructions. This explains why trends of fashion are so numerous today in the scientific field and why conceptual monocultures have such a tendency to proliferate to the detriment of approaches involving the crossfertilization of objects of research, perspectives, and points of view—in short, approaches that express true intellectual curiosity and multifaceted interests.

This prospective approach, which attempts to restore theoretical memory, is driven by a conviction: that the patterns of implantation of communications and information technologies and the development of their uses are social constructions. They take shape through adaptations, transitions, resistances, and above all through contradictory meanderings where collisions of different actors, ideas, material interests, and social projects are sure to occur. Consequently, conceptual tools forged to designate and account for these situations are open to radically different, even opposed, interpretations and uses.

Laying out the trajectories of thought on the progressive internationalization of the communication system also means investigating the conditions for realizing an international democratic space, perhaps in the form of those "networks of mutual aid" dreamed about at the end of the nineteenth century by the anarchist geographer Peter Kropotkin, who believed in the possibility of technical networks becoming tools for reinforcing community bonds.

Why this concern for, and insistence on, the need to deepen our understanding of the links between communication and space, the international and democracy, at a time when, according to some, History is finished and the system embodied by neoliberal capitalism has definitively "won"—and naturalized its conception of democracy and individual freedom? It is fitting first to recall that a hundred years ago the Victorians themselves were under the illusion that history had reached its end, since the British Empire symbolized in their eyes an unsurpassable horizon, total perfection. We may add that in the new global order, where the West continues to preach to the world about democracy, the model of development that survived the Cold War has in fact left more than three-quarters of humanity on the margins of the material benefits of market society, even while integrating them into its symbolic universe; and even in the so-called rich societies, the mechanisms of social exclusion are by no means declining. It was precisely because of the need to contest this "single" economic model that the heads of state meeting in 1992 in Rio de Janeiro at the United Nations-sponsored World Conference on Development and the Environment (the "Earth Summit") ran into difficulty; the gathering produced only the meagerest results. A "Global Forum" of nongovernmental organizations from more than 170 countries, meeting simultaneously in Rio but on the margins of the World Conference, bore testimony to the sources of worry for the various civil societies throughout the world and the need for "different models of social transformation." These first "Estates General of World Civil Society" marked an institutional and conceptual break. They signaled a gradual movement toward a new way of envisaging the construction of the social bond in its international dimension.

There, too, communications is at the heart of the search for new forms of international relations, through the rehabilitation of "social networks," an essential prerequisite for all utilization of technical networks by this new type of citizens' movement. By refusing to engender a hierarchical structure and by privileging a network mode of organization in the name of a shared if not coherent vision, this molecular action of civil society helps to reinvent an international public sphere that takes an active part in the debates on alternative economic models, "sustainable agriculture," external debt, drugs, racism, and so forth—all major themes that the egalitarian myth of the "Global Village" and the new technological order of communication tend to sweep under the carpet. The globalization of image flows has no democratic virtue in itself; it only acquires it if individual participation is not limited to the role of "voyeur" of the world and its great social imbalances.

As a critic of the French edition of this book observed: "Doubtless one may hate the 'information society,' but the issue is not to evaluate whether it is better or worse than its predecessors; neither societies of sovereignty nor societies of discipline deserve the slightest tear of regret. The issue is rather to invent the tools to combat its domination, and the means to move away from it. The author has given us a strategic map." Attempting to draw a strategic map is in fact the best definition of my purpose.

A.M.

Preface to the Original French Edition:
The Communications Triangle

The task of this book is to reconstruct the genealogy of the sphere of world communication. We will follow a multidimensional approach, analyzing the ways in which technologies and networks have taken root since the nineteenth century and have ceaselessly pushed back the frontiers of the nation-state. At the same time, we will bring to light the concepts, doctrines, theories, and controversies that have marked the construction of the scientific field known as international communications.

This history of international communications and its representations is a history of the interwoven paths of *war*, *progress*, and *culture*, and the trajectory of their successive rearrangements, their ebb and flow.

Communication serves first of all to make *war*. However, outside periods of hostilities, which foster a wealth of analyses and deploy them in the service of armies, war has traditionally constituted a blind spot in thought about communication. Confining the notion of communication to the entertainment industry in peacetime is merely the latest way of stifling examination of the relation between communication and war. Reading manuals on psychological warfare intended for the armed forces tells us more about the subject than most textbooks with which future professionals in communications learn the rudiments of their field.

War and its logics are essential components of the history of international communications and of its doctrines and theories, as well as its uses. This has been true since the advent of the telegraph and the photograph. Nearly 140 years ago, during the Crimean War, jurisprudence was established on "depicting the war," the first of a long series of decisions that culminated in January 1991 in the total control of information by military authorities. The first mass experience from which the theory of communication drew its hypotheses on the management of public opinion was World War I, a "total" conflict that affected more and more people as it went on. The notion of "propaganda" took off then, later to metamorphose into "psychological warfare" during World War II, having first been relayed by the doctrines of the Nazi party and by the *agit-prop* of the young Soviet government and the Comintern. It is highly symbolic that the first attempt to construct a discipline called international communications took place in the United States in the 1950s, during the height of the Cold War, under the wing of empirical sociology.

For more than 40 years, on both sides of the Iron Curtain, the Manichean logic of that undeclared war between East and West would furnish not only a frame of reference for international communications conceived in terms of "blocs" but also an industrial matrix for the development of high technologies of communications and information—until the day when what had been thought monolithic and established for all time showed its fissures and contradictions. The colonial wars against liberation movements would force the general staffs of the Western armies to break with doctrines of conventional war and elaborate strategies of "psychological action" inspired by the tactics of their adversaries. It would take the shock of the Gulf War to remind us that psychological war and willful and officially approved disinformation are far from having disappeared from the audiovisual landscape in the era of transnational communications. This war further served to remind us how much older methods have been modernized since the end of the wars in Southeast Asia, making ever more fragile the thin partition separating information from propaganda. Nor should we forget that the nature of war itself has changed profoundly.

Communication also serves to promote *progress*. With the prodigious takeoff of electronic and computer technologies, "communications" has become, in the 1980s, progress itself, and the rise of technical networks in the "information society" has become the measure of the growth of democracy. Since the 1970s, the utopia of communicational egalitarianism has inspired inventors of concepts such as "the global village," "the technotronic society," and others that have accompanied the progressive slide from communications seen as the activity of a particular sector to communications seen as the very foundation of the new society. Paradoxically, however, the large industrial countries were not the original theater of operations of the theories of communication-modernization-progress; it was located instead, during the 1950s and 1960s, in the planet's zones of insecurity, that is, in the countries generally referred to as the "Third World." Communication as an agent of social change was proposed there as the means of overcoming underdevelopment, making its contribution to the "rationalization" of behavior in the domains of family planning, agricultural methods, and alphabetization. Then came the crisis of the linear conception of development/progress that had been inspired by the industrial history of the West. A new field of debate opened on the conditions of international democracy in cultural exchanges, signaling the emergence of other actors, other cultures, other histories, and other ways of seeing than those consecrated by the historical experience of the masters of technology. It is not the least of the virtues of the history of international communications that it provokes a challenge to ethnocentric points of view. The public debate launched at the beginning of the 1970s was interrupted in the following decade, leaving unresolved the question of the international management of the inequality of development, a question that the crumbling of socialism, and the new agenda for integrating the former Eastern bloc into the world market, will not help put back into the foreground.

Communication is also *culture.* Placing thought about communications under the sign of culture, however, was not a major concern of the theories and strategies of international communications in the course of their history, because of their technicist and economistic drifts. It was only recently, in the 1980s, that the recentering on culture acquired its legitimacy, as centralized models of the management of culture in the welfare states entered into a crisis and as the world market became a space of transnational regulation of the relations between nations and peoples. Other factors were the critical challenge to the pseudo-universality of a model of growth and development unable to deliver to all what it has realized for a few, and the crisis of the idea of the nation, which became problematic in the eyes of many people. One culture or many cultures? This question is henceforth of central importance in the construction of a world space, a space affected at once by the deterritorializing logics of deregulation and globalization of the world economy and by the process of reterritorialization of particular spaces as each community tries to give meaning to the deep tendency that thrusts it into contact with the "universal."

Today, theorizing communication from the standpoint of the international seems essential. Knowledge of what is happening in the world space in formation is in fact indispensable for combating the many forms of exclusion of the Other. By making a detour via this dimension of reality we allow ourselves to open up a debate that, because it has remained confined to the national perspective and has failed to take the critical distance conferred by history, has become a vicious circle in which the person who puts the question is at once judge and interested party.

PART I

War

Chapter 1

The Emergence of Technical Networks

The Mastery of Space

The nineteenth century saw the slow emergence of a new mode of exchange and circulation of goods, messages, and persons, as well as a new mode of organizing production. In the course of the century, and especially after 1850, as the notion of freedom of opinion became more concrete, a variety of technical inventions made possible the development of new networks of communication. The historical forms in which each of the new circuits of exchange took root in diverse societies raises questions that resonated into the following century.

It was at the end of the eighteenth century, during the period of the Revolution, that France established the first system of telecommunication—although the term was not coined until the beginning of the twentieth century by a French engineer of the *Postes et Télégraphes,* to be ratified internationally in 1932. The basis of this system, which marked the first victory over time and space, was the optical or aerial telegraph—still referred to as the semaphore telegraph—invented by Claude Chappe. Approved by the French National Convention, the first link by overhead telegraph, which consisted of the transmission of repeated mechanical signals from post to post, was installed between Paris and Lille in 1793. A country at war needed to establish communication among its armies. The first use of this invention was thus for military ends, and so it would be used for a long time. In fact, when Chappe proposed to Consul Napoléon Bonaparte three civilian uses of the telegraph, two were refused: putting it at the disposal of industry and trade and launching a small telegraphic gazette. The only civilian use permitted was the transmission of lottery results (which foiled crafty speculators who played on

3

the delay in postal delivery between capital and province). As Yves Stour-dzé observed, "for 50 years, the optical telegraph was financed by subsidies from the Ministries of War and the Interior and by the national lottery. Only when a 'strong' state arose would the risk be taken of opening the electric telegraph to the public."[1] During these 50 years French telegraphy built up the longest network in the world, consisting of 534 semaphore stations over nearly 5,000 kilometers. Like the road network, and later the rail network, it was constructed on a star system outward from Paris. The capital was in direct correspondence with 29 cities, with two outposts at Mainz and Turin.

France opted for state monopoly. But when the United States inaugurated its system of semaphore telegraph in 1800, it adopted the commercial model. In England, between 1797 and 1808, the Admiralty built links between London and four coastal ports. After the Napoleonic Wars, a permanent network of semaphores was installed for military needs, while the lines destined for use by shippers and traders were built and operated by private firms. In Prussia, on the other hand, use of this technique, established only in 1832, would be the exclusive domain of the general staff.[2]

The preelectric telegraph aided the French state, as it emerged from the Revolution of 1789, in its project of mastering space. It was an element in a unified territorial scheme. A coherent vision of the national territory gave form to regulations assuring the flow of merchandise and people. Barriers between provinces were abolished, administrative divisions were stabilized; the tax system and the legal code were unified; French was imposed as the language of the nation-state. A corps of "national engineers" was created, combining military engineering and public works (bridges and roads, ports, canals, and later, railways); the École Polytechnique was formed in 1795. A series of indispensable measures for economic and demographic management were passed. From this period date two pioneering "normalizing" initiatives.

In 1792, a commission of weights and measures began to work on the definition of the metric system. Seven years later, a law endorsed the conclusions of this scientific assembly on the new basic units, putting an end to the patchwork of weights and measures in use in various regions of the country (although it was not until 1840 that the meter, the gram, the liter, and the hectare became mandatory). The norm was already becoming internationalized: following a number of European countries, Germany settled on this nomenclature in 1872, and Switzerland in 1877. A hundred years after the French Revolution, England, Russia, Japan, and the United States would continue to make use officially of nonmetric measures, but most of Latin America rallied to the metric system.

A second administrative innovation occurred in 1796, when the Ministry of the Interior inaugurated the first official bureau of statistics, that is, the first state institution responsible not only for collecting and classifying reports and documents communicated to government but also for directly organizing the recording of facts for a general statistics. Bavaria, Russia, Italy, and Prussia took this step between 1800 and 1805. But it was not until 1832 that England acquired the same tool—a year later than Belgium, which, under the inspiration of mathematician and astronomer Adolphe Quételet, rapidly became the model for other countries in establishing statistical services. (At the beginning of the nineteenth century, the practice of taking a periodic census of inhabitants also began to be institutionalized in England and France in 1801 and in Prussia in 1810. Before that, only the Scandinavian countries—for example, Sweden in 1748—and in 1790 the United States, had carried out such operations.) The formation of statistical offices and the emergence with Quételet of a "political arithmetic" gave rise to the first suspicion of the disciplinary functions of these methods of managing numbers. In deducing averages from his statistical series, Quételet proposed norms, an "average individual" desirable for social equilibrium and a "moral order."

The setting up of a service of general statistics on the initiative of the French Ministry of the Interior at the end of the eighteenth century was contemporaneous with another project launched by the same institution: the organization of an annual exposition of the products of French industry. The aim of this exhibition was to give an overview of national production and technical and scientific innovations, with an eye to "stimulating French industrialists in the fight against monarchic England." The first exhibition took place in Paris in 1798, and there were ten more in the first half of the following century. A jury composed of eminent scientists designated by the government awarded prizes to companies, inventors, experts, engineers, artisans, and workers who distinguished themselves by the quality of their products or works. The need to systematize the order in which the products were displayed coincided with the concerns of experts responsible for developing a nomenclature for collecting and interpreting data on the main trends in industry.

But for a historian of communication, the expositions organized in this period take on a more precise interest: they initiate a new form of communication by placing science, industry, scientific research, and technical innovation on display. It is by means of the symbolic values they offered their visitors that these events progressively constructed the grand narratives of civilization-as-progress. These grand narratives, this progressive utopia, would not unfold in total freedom, however, until the expositions stopped being confined to national products and achievements and became showcases for the whole world. This would happen in the universal expositions of the second half of the nineteenth century.

The Specter of the "Black Cabinet"

Who should control the circulation of information, the installation and functioning of long-distance communication networks—the state or the private sector? Who should be authorized to use the new services? These questions predate the arrival of the manual telegraph. They were posed during the long history of postal institutions.

The Renaissance and the generalized use of paper converted the post into a regular service. One of the first initiatives was taken in 1464 by the French king Louis XI. Inspired by the organizational model of the Roman post—which in turn had adapted that of Cyrus the Great's Persia (529 B.C.)—and by the network established in 1150 by the University of Paris for its own purposes, the edict of Luxiès instituted postmasters throughout French territory and laid the foundations of the system: relay stations, the essential links in the chain of postal delivery. The king of England would do the same in 1481, when vassals were authorized to use the king's couriers. But the king reserved the right to read any letters entrusted to them.

It was more than a century later, under Henri IV, that the French state took regular charge of the delivery of correspondence between individuals, previously the domain of private initiative. It was mandatory to keep records of names and addresses of senders, both for reasons of state security and because of the mechanisms of a service that was not paid for until a letter reached its destination. From this era onward, the history of the *ancien régime* of the post merges with the progressive unification of services into a single monopoly for the transport of mail, a monopoly that sealed the fate of a system in which various agencies, including the university, shared control. In the construction of this new administrative model, one name stands out: Louvois, the postal superintendent-general from 1668 to 1691.[3] This minister of Louis XIV has gone down in history not so much for having reorganized the postal system as for having radically transformed the military system. Louvois paved the way for a modern army by substituting a permanent and regular army for improvised troops, reforming discipline, and organizing a corps of engineers. In suppressing private offices in 1635, England preceded France by very little in the consecration of the post as a royal prerogative.

In a postal landscape dominated by the direct control of the sovereign, one major exception existed: the territories belonging to the Holy Roman Empire.[4] In about 1450, the German emperor Frederick III confided to the patrician Thurn and Taxis family from Venetia the organization of a postal service for a broad public. This mission was confirmed in 1516 by Charles V, who, at a time when the Reformation was reaching the smallest villages, counted on the post to assure—from a central position in Brussels—a permanent link between central power and the farthest-flung subjects of the Empire. More than one large Hanseatic city seeking to assume its own control over mail-carrying would rebel against this privilege. Despite changes of regime, invasions, and war, the Thurn and Taxis postmasters kept a private monopoly for over three centuries on what was in effect the first trans-European network to defy territorial sovereigns. It was not until 1867, three years before the political unification of Germany, that the family lost its last monopoly over a postal service, with the merging of the post and telegraph over a major portion of German territory into a single state administration.

Throughout the prehistory of postal modernity, the fear of national or international conspiracy obsessed the directors of postal networks. This resulted in France during the reign of Louis XIII in the "Black Cabinet," a bureau where postal secrecy was violated. The revolutionaries of 1789 denounced this practice in their lists of grievances as "one of the most absurd and infamous inventions of despotism." A British historian goes so far as to say of the development of the post in his country: "Anyone writing the early history of our Intelligence Service would at the same time be writing an account of the beginning of our postal services."[5]

The equivalent of the French "Black Cabinet" would come into existence in several countries and would stay active long after official acknowledgment of the citizen's right to confidential correspondence. Although the French Revolution abolished it, Napoleon reestablished it shortly afterward and extended the scope of his activities. If we may believe a former censor of the Russian czars' Black Cabinet, clearly proud of his job: "It is fair to say that nowhere in the world has a 'black cabinet' worked as well as that of Russia, particularly that of St. Petersburg."[6] On the eve of the 1917 Revolution, a Black Cabinet still functioned in the Russian Empire. The Soviet Union, after having held it up to public opprobrium, quickly reestablished it with formidable efficacy.

The Telegraphic Vanguard

In 1837 a pair of Englishmen, William Cooke and Charles Wheatstone, and an American, Samuel Morse, perfected in their respective countries the first systems of electric telegraphy. The first client, starting in 1839, was the railway system in the suburbs of London. (Up to then the system, established in 1825, had used hand signals to communicate from station to station.) Private companies were in charge of management. By 1852, England, already endowed with the densest rail network in the world, could boast 6,500 kilometers [4,000 miles] of telegraphic line. She was ahead of the United States, which established the first intercity line (Washington-Baltimore) in 1844. The shift to commercial exploitation of the telegraph took place at the end of the Civil War in 1865 with the founding of the Western Union company, which unified the U.S. network.

France was one of the last countries to adopt the Morse system, since it preferred an electric version of the Chappe, known as the Foy-Bréguet. Although the decision to establish a telegraph network was made in 1845, the project was held up for an additional 7 years, picking up speed only in the 20 years that followed. In the same lapse of time, the rail network grew from 3,010 to 17,733 kilometers. In 1852 the use of telegraph by the public was finally authorized, but the new regulation was full of restrictions, in the name of public order and decency. We find again the same fear of conspiracy, combined with a dread of illicit speculation. No real liberalization took place until after 1867.

Meanwhile, the use of telegraphy for strategic purposes was tested on a number of occasions. The overhead telegraph installed in Algiers in 1842 proved a decisive aid during the occupation and colonization of Algeria, and was relayed after 1854 by the first electric telegraph lines. But the first ambitious application of this system by the general staff took place during the Crimean War. Fighting against the czar's army, the French and British high commands established a liaison between headquarters and the various army corps in a coalition that also included expeditionary forces from the Piedmont and Turkey. The British Royal Engineers also laid an underwater cable in the Black Sea to assure a permanent link with Paris and London for the duration of the conflict. The Civil War in the United States stimulated the construction of 24,150 kilometers [15,100 miles] of cable in 4 years and the sending of more than 6.5 million telegrams.

In addition, the Crimean War allowed the military authorities to lay the foundation of new jurisprudence in the area of wartime censorship. For the first time in the history of modern media, images of the theater of operations were censored.[7]

The British photographer Roger Fenton was authorized to take pictures on the condition that his lens carefully avoid capturing the horrors of the war. "So as not to frighten the families of soldiers" was the reason given by the general staff. The result was 360 plates in which the war appeared as a picnic. This artificial construction contrasted markedly with the reports of journalist William Howard Russell, who in the London

Times described the slaughter in Balaclava on October 25, 1854, in which 400 of the 600 British cavalry sent up against the Russian cannons perished,[8] giving rise to an unprecedented awakening of public and parliamentary opinion. In February 1856, one month before the signing of the peace treaty, the British high command decided to put a stop to the free exercise of journalism. A decree made accreditation by the military authorities mandatory.

Less than ten years later, photographer Matthew Brady brought back from the Civil War thousands of daguerrotypes that were not subjected to advance censorship: scorched earth, burned-down houses, families in distress, corpses. Sales did not match his expectations, and Brady lost his fortune in the venture.

The Penny Post

As an indication of the frontier reached by long-distance communications up to that point, the International Telegraph Conference, organized in Rome in 1872 in order to regulate the new networks on the planet, brought together only 22 states: Germany, Austria-Hungary, France, Great Britain, the British Indies, Italy, Russia, Turkey, Spain, Belgium, Holland, the Dutch Indies, Romania, Sweden, Denmark, Norway, Switzerland, Greece, Portugal, Serbia, Luxembourg, and Persia. All these countries or territories had joined the International Telegraph Union, founded at a first congress in Paris in 1865. The contracting parties had granted any person the "right to correspond by means of telegraph," reserving the right to prevent the transmission of private telegrams "that threaten state security or violate the laws of country, public order or morals." This Union represented the first international institution of the modern era and the first organization for the international regulation of a technical network.

The universal Postal Union was created a few years later, in 1875 in Bern, under the Universal Postal Convention of 1874. A first postal conference grouping together representatives of fourteen nations had been held in Paris in 1863. The initial decisions of this union were to harmonize international postal rates and to recognize the principle of respect for the secrecy of correspondence. The dusting off of old postal institutions was the order of the day.

England gave the example in 1840 with the Rowland-Hill reform, named after its initiator, who was also the inventor of the postage stamp. The novelty was not to take account of the distance in the price for carrying postal correspondence, and to adopt a single rate of one penny. As a result of this measure, the number of letters jumped from 76 million in

1839 to 169 million the following year, an increase of 122 percent. The financial consequences were less fortunate, and it took 23 years for the British Post to surpass its net revenue of 1839. But as the French economist Paul Leroy-Beaulieu noted at the end of the last century: "The low price of mail helped the development of trade, facilitated the growth of industries which, under the former rates, would have remained stagnant, and by thousands of indirect channels the Treasury received, through the increase in the revenue raised by other taxes, a sum that almost cancelled out or at least attenuated the loss it suffered by the long decline of postal revenue." He concluded: "The postal reform has been, without a doubt, along with railways, steamships, the telegraph, and Australian and Californian gold, among the powerful and diverse causes of the magnificent growth in trade in the past 35 years."[9]

In 1847, the United States Congress adopted the postage stamp and copied several measures from the British reforms. The following year, France issued its first stamp and engaged in a less radical reform than the British one. The administration set the domestic rate at 20 centimes for letters weighing up to 7.5 grams. The increase in postal traffic was much smaller, a mere 34 percent.

While in the United States the telegraph remained under private control and mail service was public, the 1860s saw the merger throughout Europe, under government administration, of post and telegraph: in England in 1868, with the purchase of the telegraph network from private companies; in France in 1878; in Italy in 1889, and so on. The governmental status of post and telegraph remained diverse: in some cases they were given their own ministry, but in others they were attached to an existing ministry. At the end of the century, the two services were placed under the Ministry of the Interior in Norway, Romania, Russia, Spain, Greece, and Mexico, and under the Ministry of Foreign Affairs in Bulgaria; certain states preferred to make them dependent on the Treasury. Such was the case in France at times, and in Sweden. Metropolitan models of organization of the postal service were exported to their peripheries. Thus Siam hired a German expert, and Persia an Austrian. Japan brought in an American when the feudal regime was abolished (1871), thrusting the country into the era of networks: the first telegraphic line went into service in 1869–70 and the first railroad in 1872, the year when the emperor undertook the unification and modernization of the army, and one year before the adoption of the Western calendar.

With the improvement of steam navigation, the transport of international correspondence was remodeled. Nevertheless, it was not until April 1838 that a steamship first crossed the Atlantic and entered the port of New York: the *Sirius*, belonging to an Irish shipper in Cork. In 1839, the

British Admiralty accepted an offer from Samuel Cunard, backed by cotton importers, to establish a bimonthly service between Liverpool, Halifax, and Boston. The Admiralty, wanting to strengthen the imperial link with Canada, which had rebelled in 1837, granted this first postal company in the North Atlantic an annual subsidy of 35,000 pounds. In 1840, another postal contract was given to the Royal West India Mail Steam Packet Co., for the delivery of mail to the Caribbean and Brazil. France did not enter the race for international postal services by steamship until 1854, with the creation of the Compagnie Générale Maritime, ancestor of the French Line. Its founder was Emile Péreire, a Saint-Simonian and project manager of the first French railway, Paris to Saint-Germain (1837); its first president was Adolphe d'Eichtal, also a Saint-Simonian and president of the same railway.[10] In most countries, the system of subsidies to companies engaged in shipping mail was the rule at the dawn of the twentieth century. None of the lines assuring this service was able to survive without state support, either in France or in the United States or even in England, despite the powerful resources of that country's commercial fleet.

All these technical upheavals in modes of communication contributed to a radical change in the economic status of information. As the time necessary for transmitting messages diminished, modifications became necessary in the methods for collecting, treating, and codifying information. The existing means of regulation of the stock markets became archaic and new procedures for intervening in the market had to be found. Information became a matter for specialists and its complexity called for the skills of forecasters. How did businessmen of the time experience these upheavals? One answer is provided in a financial newspaper, *The Sugar Cane*, of Manchester, in 1888: "In the good old days, commodities rarely produced losses, except in times of great panic. Tradesmen, even when they speculated far and wide, had existing commodities in warehouses and ports close at hand. Prudence, prediction, and intelligence were rewarded. The introduction of steamships changed all that, and the telegraph has completed the revolution. Exclusive information, laboriously acquired, which formerly brought an intelligent merchant a profit, has today become public property from the moment it appears. It is now available at literally the same time to the audacious speculator as to his competitors; this is now the rule."[11] An indication of the increasing flow: according to statistics compiled by the International Telegraph Union, the number of telegraphic transmissions in the world soared from 29 million in 1868 to 121 million in 1880. Twenty years later, they reached 329 million. The international flow represented, at this point, slightly more than 20 percent of the total.

Since 1875 the planet had been in the period characterized by the

British historian Eric Hobsbawm as the "Age of Empire" (1875–1914). With the acceleration of technical changes and the takeoff of the colonizing countries, the chasm was dug between the "developed world" and what would later be known as the "Third World." In 1800, the North-South gap between gross national products per capita was insignificant; it was 2 to 1 in 1880, but it would be 3 to 1 in 1913 and 7 to 1 in 1970.[12]

The Wire Era

In the United States in 1876 Alexander Graham Bell patented the invention of the telephone. The following year, the Bell Telephone Company undertook the commercial exploitation of his machine. In 1882, Bell set up its first subsidiary, at Anvers in Belgium. In 1885, American Telephone & Telegraph (AT&T), later to become the head office of the Bell System, was founded. For over 80 years, AT&T succeeded in keeping its near-monopoly over the telecommunications networks of the United States. It would take the groundswell of deregulation at the beginning of the 1980s to dismantle it.

In 1881, the American network had 123,000 telephones. In Europe, Great Britain first chose to let the private sector carry out the extension of lines, but London had only 1,100 subscribers in 1881. Not until 1912 did the state take control of the entire system. In France, the telephone followed the "administrative model" which had already proven its worth for the telegraph. But, as across the Channel, administration fell behind. In 1888, Paris had 5,800 telephones, or 70 percent of the total in the ten largest French cities. The first international telephone calls were exchanged between Paris and Brussels in 1887, between London and Paris in 1891, and between Paris and Switzerland in 1892. On each occasion, the signing of a bilateral agreement was necessary to authorize transmission. It was not until 1906 that the first multilateral agreements were signed, at the Berlin Conference on radiotelegraphy that saw the birth of the International Radiotelegraph Union. It must not be forgotten that at that time the area covered by international telephone lines was relatively limited. Telephone networks did not acquire a world dimension until 1956 when the first telephone cable was laid under the Atlantic—only a year before the satellite race broke out.

By 1900, the gaps within Europe were already noticeable. While Sweden had one telephone for every 115 people, and Germany one for every 397, France, with only one telephone for every 1,216 people, surpassed only Italy (one per 2,629) and Russia (one per 7,000). This lag with respect to other countries would remain a permanent structural feature of France's telecommunications practically until the arrival of the digital phone after 1974. (In less than 15 years, the number of telephones more than quadrupled.) At the end of the nineteenth century, the United States was well in the lead, with a ratio of one telephone per every 60 inhabitants. It was the U.S. manufacturers of telephone equipment who wove the first multinational network of production and sales. International Western Electric, subsidiary of Western Electric, itself owned by AT&T, set up branches in Great Britain, Belgium, Spain, France, Holland, Italy, Norway, Poland, Australia, China, and Japan.[13] (In 1925, after an antitrust suit, it would cede the network to International Telephone and Telegraph—ITT—founded in 1920, and would not again gain a foothold abroad until after 1982, as a consequence of deregulation.)

In 1901, the Italian Guglielmo Marconi exploited the discovery of the propagation of electromagnetic waves and, with the support of naval armament companies and of newspaper groups, succeeded in establishing the first wireless trans-Atlantic telegraph transmission. The British Admiralty was the first to grasp the strategic implications of this innovation in radio communications, which pushed back the limits of communication on a globe already profoundly changed by underwater telegraphic cables. The Italian inventor's patents served to found the British firm bearing his name. In 1899 the French War Ministry commissioned Captain Ferrié, author of the first book on radio technology, to write a study of its military applications but asked him to avoid contact with Marconi, who was "in the service of a foreign power." In 1907, the French Navy took the initiative of grouping together the first companies in this field.[14] At the Berlin Conference in 1906, 28 states had debated equipment standards and procedures to minimize interference between stations, and the great naval powers, major users of radio (Great Britain, Germany, France, the United States, and Russia) had legitimated an imperial doctrine of radio frequency allocation, allowing priority to the country that first notified the International Radiotelegraph Union of its intention to use a specific radio frequency.

August 18, 1858, seven years after the laying of the underwater telegraphic cable linking Dover and Calais, the first link between England and the United States was established. Nevertheless, it was not until September 1866 that the trans-Atlantic cable became really operational. The first commercial message was a dispatch to the *New York Herald* transcribing in its entirety the speech of Emperor William, victor over the Austrians at Sadowa, before the Prussian Parliament. But the cost was prohibitive.

The international extension of cable was marked by the rivalry between the British and French empires, which intensified after 1869, with the opening of the Suez Canal. (Between 1851 and 1868, underwater networks developed essentially through links in the North Atlantic, the Mediterranean, the Indian Ocean, and the Persian Gulf.) The years 1870 to 1880 saw the successive inaugurations of communication links between the English coast and the Dutch East Indies (Batavia), the Caribbean network, the line from the British West Indies to Australia and China, the networks in the China and Japanese seas, the cable from Suez to Aden, communication between Aden and British India, the New Zealand cables, communication between the east and south coasts of Africa, and the cable from Hong Kong to Manila. In the 1880s, France established a series of links along the coast of Indochina and black Africa (with networks in Senegal and on the west coast of Africa). Another direction in which the network centers of Europe and the United States moved was toward Central and South America: in 1874, the south trans-Atlantic network opened, linking Lisbon with Recife, Brazil, via the Cape Verde Islands and Madeira; in 1876, a network was established along the coast of Chile; in 1880, another began along the coast of Mexico, and a year later the networks along the Pacific coast from Mexico to Peru were in operation.[15] These cable networks that spanned the globe were largely in the hands of the private sector. Of the total cable distance of 104,000 miles, not more than 10 percent was administered by governments.

British supremacy over the underwater networks was overwhelming: in 1910, the Empire controlled about half the world total, or 260,000 kilometers. France, which in contrast to the United States and Great Britain opted for state administration of cable, controlled no more than 44,000 kilometers.[16]

The preponderance of British companies, which lasted until the end of World War I, rested on a dual control of international networks: direct control through ownership, and indirect control by means of diplomatic censorship, which London exercised over the messages traveling through British channels. From an industrial point of view, the Victorian Empire combined technology, factories, and a cable fleet, all of these more efficient than those of its rivals. It controlled the copper and guttapercha markets, since the world rates of these two raw materials for the manufacture of cable were fixed in London; British mining enterprises owned copper deposits and mines in Chile, the world's biggest producer. Finally, the Crown spared no effort to assist these companies, either scientifically (Admiralty researchers and the cartographic service) or financially (higher subsidies than those of France). In 1904, 22 of the 25 companies that managed international networks were affiliates of British firms; Great Britain deployed 25 ships totaling 70,000 tons, while the six vessels of the French cable fleet amounted to only 7,000 tons.[17]

In 1898, during the Fashoda crisis, the culmination of the dispute between the two great colonial powers, when the French expansion plan from west to east, starting from Brazzaville, collided with Britain's north-to-south plans, Paris was dependent on its rival's networks, and had to ask the government in London for authorization to use the cable and General Kitchener's ship to communicate with Captain Marchand, who had just occupied Fashoda. The Havas news agency found itself in a similar situation of subordination with respect to the Reuters Agency at the time of these events, which, as historians of the press note, saw the recognition of "the weight of public opinion in international relations, such as the popular press reflects and refashions it."[18]

First Genres of Mass Culture

Very early on, the large news agencies became assiduous users of long-distance communications, happy to depend no longer on carrier pigeons to transport their dispatches.

The French Havas Agency (ancestor of AFP, Agence France Presse) was founded in 1835, the German agency Wolff in 1849, and the British Reuters in 1851. The American agency Associated Press (AP) began its history in 1848. But only the three European agencies began as international ones; not until the turn of the century did the American agency move in this direction. As a cartel, Havas, Reuters, and Wolff divided up the world market in an explicit agreement signed in 1870. The special territory of the Parisian agency was Mediterranean Europe, while that of Wolff was central and northern Europe. Reuters, meanwhile, followed the geographical lines of the British Empire. From the start, it made commercial and financial information its privileged area. Havas's originality was to combine information and advertising. This plural function made it the precursor of the multimedia groups of the twentieth century. After World War I, Wolff ceased to be a world agency and Havas and Reuters emerged strengthened, until the 1930s when the U.S. agencies AP and United Press (UP) began to hunt for news on their terrain.[19]

The rise of the large agency networks paralleled the advent of a press freed from the constraints of censorship. From 1853 to 1861, Great Britain eliminated all the "taxes on knowledge" that had hampered the development of a mass press. The United States had preceded it in this, since even before 1850 an inexpensive daily press with a popular readership had appeared there.

In France, landmark legislation in 1881 freed printing, newspaper stands, and bookshops. Prior censorship was abolished, as was the surety bond and the stamp tax. The only press misdemeanors were provoking crime, inciting military disobedience, insulting the president of the Republic, sedition, indecency, defamation and personal injury, and offending heads of state and foreign diplomats. Newspaper peddling, sales, and billposting were permitted. The publisher was to be responsible for the publication, and his name was to appear on the masthead. These were the terms of the French law of July 29, 1881, hailed as the great law on the freedom of the press and considered a victory for the Republican bourgeoisie.

In 1890, *Le Petit Parisien* prided itself on being the first popular daily paper in Europe whose circulation exceeded one million.[20] William Randolph Hearst's *New York Journal*, symbol of the sensational press, could not reach this figure, despite its Sunday supplements and its comics. On both sides of the Atlantic, competition stimulated search for the first genres of mass culture. In France, where *Le Petit Journal* and *Le Petit Parisien* were engaged in a fierce battle, the serial story (*feuilleton*) became one of the trump cards of popular journalism. Introduced in 1836, this genre reached its apogee in the mid-1880s, when papers published two or three stories at a time amid large promotional campaigns.

In the United States, the fight between the Sunday supplements of Hearst's paper and Joseph Pulitzer's *New York World* gave rise to the first comics in 1894. Less than 15 years later came the first strategy for penetrating the international market using this kind of editorial product. In 1909 Hearst created the first syndicate, International News Service, an agency whose function was to sell literary material, articles of popular science, crossword puzzles, and comic strips to newspapers. It was succeeded in 1915 by the King Feature Syndicate, one of whose staple products was comics. A consequence of the restructuring of this genre to meet the needs of the syndicate was the end of a craft style of production in favor of a more industrial division of labor (the agency retained the copyright and could retouch, eliminate, and modify the material, find a successor when a cartoonist left, and generally hold editorial control). Further, there was, in the words of Roman Gubern, a "standardization of material, which brought about a certain homogeneity in the international market and eliminated critical or aggressive aspects that might alienate customers in countries with different customs, religions, or political principles."[21]

But it was with the film industry that the first major internationalization of nascent mass culture began. The first film screenings took place in Paris and Berlin in 1895, and the following year in London, Brussels, and New York. The Lumière brothers disputed with Edison the claim to invention of this technology. The rivalry between the two systems was particularly felt in the large Latin American cities, which discovered cinema at the same time as the major European capitals. In fact, the first public projections in Latin America took place in 1896 in Rio de Janeiro, Montevideo, Buenos Aires, Mexico City, Santiago, and Guatemala City, followed in 1897 by Havana and Lima. In Mexico and Brazil, the large provincial capitals—Guadalajara, Mérida, Puebla, Curitiba, São Paulo, Salvador de Bahia—soon saw the arrival of the film projector, whether that of Edison or the Lumière brothers. But the capital of Bolivia, La Paz, had to wait until 1904 for a first screening; Bolivia's mining center, Oruro, saw no film until 1907.[22] After the phase of traveling fairs and exhibitions, where spectators had to stand up, the cinema became a sedentary pleasure in 1902–3 in the United States, and three or four years later in France and Germany. On the eve of World War I, it was French producers, with the Pathé brothers (their firm was founded in 1907) at their head, followed by Gaumont (1885), who clearly dominated the European market. Pathé's distribution bureaus were located in ten countries (Germany, Italy, Spain, Switzerland, Holland, Portugal, Sweden, Turkey, the United States, and Brazil).

The Hollywood film industry was initiated by the independent studios between 1909 and 1913. In 1910 Carl Laemmle launched the first star, Mary Pickford, inaugurating the star system. In 1915 the Hollywood studio Reliance-Majestic produced *Birth of a Nation* by D. W. Griffith, a landmark film about which Jean-Luc Godard reflected: "The great national cinemas have always been marked by war films, especially civil war films, in other words by moments when a nation is fighting against itself and no longer knows what it is."[23]

The international trade in films was not hampered by a customs or trade policy. The first measure to oppose free trade was taken in 1916 and confirmed in 1917, when Germany imposed controls on the importation of foreign films. At the time Berlin was beginning to conceive a state policy with respect to its film industry.[24]

Creating the Event, or the Beginning of a Legend

It was in an atmosphere of bitter competition between two giants of the large-circulation daily press that a legend was forged about the power of the mass media and their relation to war.

At the end of the nineteenth century occurred the first great press campaign aimed at inciting a government to intervene militarily on foreign soil. The territory was Cuba, one of the last possessions of the moribund Spanish empire. President William McKinley proved unable to resist the pressure of public opinion stirred up by William Randolph Hearst. The Marines who landed on the island in 1898—a year decidedly rich in events, with the Franco-British confrontation in Fashoda over the carving

up of Africa and the Dreyfus Affair in France also in the headlines—
brought with them Vitagraph cameramen who filmed a military intervention for the first time, calling their report *Fighting with Our Boys in
Cuba*. This intervention, in the eyes of many historians, could perfectly
well have been avoided had not the war hysteria arisen, stoked by a press
that did not shrink from lying to provoke the fateful outcome.[25] A famous anecdote sums up this thunderous operation. Hearst sends to Havana a reporter and a well-known artist, Frederic Remington, who
telegraphed his boss from the Cuban capital: "Everything is quiet. There
is no trouble here. There will be no war. Wish to return." Which brought
the famous reply from Hearst: "Please remain. You furnish the pictures
and I'll provide the war."[26]

This incident—immortalized in the newsreel sequence of Orson
Welles's film *Citizen Kane* (1941)—counted more than a little in the
emergence of the idea that the media have unlimited power and are capable of making and unmaking governments. It also testified to two new realities. First, the use of force represented by this landing signaled the beginning of a long history of imperial policies of interference of a nature
different from customary colonial policy. Four years later, the French
company that had begun work on the Panama Canal in 1881 ceded its
property and rights to the United States, which helped the separatists in
what was still a province of Colombia to secede and proclaim independence. The new nation-state of Panama forfeited sovereignty over the
Canal Zone to the United States. The Canal opened to navigation in
1914. Mass information was already becoming an important stake. Determining whether the Spanish-American War would have taken place
with or without a strident press campaign is a question of little interest
for a prospective examination of the role of the media in society, until
one stops thinking of the media as a new demiurgic or Machiavellian
force and breaks the vicious circle of cause and effect.

It was in this region of the Americas and in this environment that the
first multinational corporations were consolidated. In their front line was
the monopoly on the planting of and trade in bananas held by the United
Fruit Company, founded in 1899, after the takeover of its competitors.
One cannot ignore this company if one seeks to retrace the genealogy of
the railroads, shipping lines, telegraph, and telephone in this part of the
world. Weaving together, or even merging, business interests and interests of nations where it held land, the United Fruit holding company—
which would pioneer in introducing the radio even within the United
States—created a subsidiary, the Tropical Radio Telegraph Company,
which operated a network of telegraphic stations in nearly 20 countries
in the Americas. Using this network to maintain links between produc-

tion enclaves as well as with the market, the company also offered tele-graphic services and railroad and steamship transport to outside clients, substituting for, complementing, or competing with, as necessary, the public communication and transport services of complicit governments that were only too happy to see what they hailed as "progress" arrive on their territory.[27]

The Iron Horse

The International Railway Conference was created in 1882, 17 years after the first international meetings concerning the telegraph. And yet Stephenson's "Rocket," the very prototype of all steam locomotives, had appeared in 1829 and the world's rail network had already reached 430,000 kilometers. The standardization of gauge came slowly. Even in England, the norm chosen by George Stephenson in 1825 for the railway line from Stockton to Darlington (4'8" or 1.435 meters), was not estab-lished definitively until 1892, despite the fact that Parliament in 1846 had granted its preference to this gauge, which coincided with that of road ve-hicles of the period. Among European countries, there was scarcely any unification of gauges. In 1844, under the influence of English engineers, Spain adopted a gauge of 1.674 and Russia chose 1.52 meters. For rea-sons of state security, neither of these two nations later sought to link up with other European networks, where the Stephenson standard had final-ly triumphed.[28]

When the representatives of the governments of Germany, Austria, Hungary, Belgium, France, Italy, Luxembourg, Holland, Russia, and Switzerland began to concert in 1882, they were not yet ready to discuss the harmonization of laws on commercial transport. Since goods were transported successively by several companies, each subject to different legislation, the problem was to determine which law should be applicable in case of loss or damage, which court would be competent to judge, and how decisions should be enforced.[29]

The reason for the relative delay in standardizing rail policies was that in contrast to the telegraph, which foreshadowed the real-time circula-tion of information in a global economy representing movement, the net-works traced by the locomotive as a machine in movement recognized the rigidity of borders, the partitions of an age in which the "nation" was the motor-force. The fate of the train, symbol of the "*dromocratic* [or speed] revolution in transport," in Paul Virilio's phrase, was tied up with the construction of the industrial nation-state and the national bour-geoisies.[30] It is revealing that the first measure taken in the United States to limit the principle of free enterprise was the Interstate Commerce Act of 1887. The United States—like England and contrary to France, which had opted for a mixed system—had left completely to private initiative the task of establishing routes and to free competition that of laying down the lines. The Interstate Commerce Act in 1887—like the Sherman Antitrust Act, passed at roughly the same time—was intended to regulate the rail network in order to allow the country to take a "great leap forward" in the Industrial Revolution, which at this stage was still necessarily national in character. The first intercontinental train had been in operation since 1869. The law would not really be questioned until the Stagers Rail Act in 1980, when the process of deregulation of the whole range of communication networks—roads, airways, rail, media, telecommunications, and also those of finance—would set on fire an economy that was in the process of becoming global.

When the "international connection" of railroads made a qualitative leap in the last quarter of the nineteenth century and the beginning of the twentieth, the process favored new imperial strategies. The formation of three companies in India after 1845 (East Indian Railway, Great Indian Peninsula Railway, Madras Railways) had already set the tone: most of the lines had been conceived for ends more strategic than economic, that is, to shorten the transport time of troops. It was not until after the outbreak of the Mutiny in 1857 that a massive railway program began, facilitating the cotton- and jute-processing industries and connecting the centers of heavy industry with the Indian coal deposits developed by the East Indian Railway. But the network of what one civil servant called "the cyclopean forges of the railways" was never completed. It did not join Bengal to Burma, which was the "rice bowl" of India. The Colonial Office deprived the country of this external outlet in order better to maintain control. "The train," declared Cecil Rhodes, "is an instrument of pacification which costs less than the cannon and carries farther." As French historian Marc Ferro reminds us in an issue of the journal *Traverses* devoted to rail networks, the train has generated its own international imaginary, not to be confused with the magical images of the Orient Express:

> Its black smoke was the very sign of progress. Through the train,
> the West was identified with its symbol. In the name of the train,
> Victoria conquered Africa from the Cape to Cairo. In the name of
> the train, Nicholas and Alexander threatened Asia and would
> reach Vladivostok. . . . The French republican rooster aspired to
> the same dream: to span a continent. Alas, between Dakar and Dji-
> bouti, its hopes were quashed at Fashoda. To dominate the empire
> of the Rising Sun, or at least obtain an audience, Uncle Sam offered
> the Mikado a small mechanical train. Just as shamelessly, Tseu-

Hi's China offered itself to the great powers in return for an express that would stop at the gates of Canton. Thus humanity in all
its colors submitted to the masters of the smoking machine.[31]

The railway model of penetration developed in the colonial context
through the exemplary network of railroads across Africa. These networks were intended to link administrative centers, located for the most
part along the coast, with mines in the hinterland, or else to allow access
to other territories to be conquered and colonized. Nine different types of
track were built, from the so-called "imperial gauge" of the Victorian era
(3'6", or 1.067 meters) to the broad gauge used almost exclusively in the
mining region of South Africa (1.435 meters). Some countries colonized
successively by two different powers even inherited a hybrid system: for
example, Tanganyika (now Tanzania) was occupied by the Germans,
whose gauge was one meter, and then by the British.

This lack of symmetry between rail systems also affected the independent nations of Latin America. The case of Argentina, with its three types
of gauge, is instructive. The most common one (1.674 meters) corresponded to the norm chosen by English engineers who had imported
equipment that had served in the Crimean War; a second (one meter) was
that of French companies, and a third, which covered only a tenth of the
network, conformed to Stephenson's gauge. This case illustrates how the
history of the railway was intimately linked to British economic hegemony in the region, at least until World War I. This is also demonstrated by
the topography of the first railway lines in Chile, which connected ports
with the British nitrate and copper mines. And since a communication
network never arrived alone, the British predominance already noted in
underwater cable combined with another, that of telegraphs and telephones. Brazil was another case entirely: on the eve of the twentieth century, there were no less than five independent railroads. Each regional
network spread out in a fan pattern into the interior from a port. The
noninterconnected regional model—with a strong concentration at the
pole constituted by the contiguity of the states of Rio de Janeiro, São
Paulo, and Minas Gerais—also governed the telegraph and telephone. (It
would prevail again with the beginnings of radio and television. In fact,
not until the promulgation, under pressure from the army, of the first
telecommunications code in 1962 and the nationalization of foreign telephone companies, did "communication" become synonymous with "national integration.")[32]

In 1870, Europe's share in the world rail network was about 50 percent; the United States was close behind. Fifteen years later, the United

States, with 365,000 kilometers, had moved ahead of Europe (202,000), Asia (24,000), Oceania (14,000), and Africa (about 7,000).

By this time, the train had already completely transformed military strategy. Used in a military campaign for the first time in 1848 (in Schleswig-Holstein), again in the Crimean War (1854–56) and in the Italian campaign launched by Emperor Napoléon III (1859), the train definitively proved its strategic importance during the Civil War in the United States. During this war, which also saw the birth of devices such as the torpedo, the Union General McClellan employed, for the first time in military history, a corps specialized in the construction and destruction of railway lines, and put into action the "Iron Horse," an ancestor of the tank.

Between 1865 and 1875, the general staffs of the German, British, and French armies, in their turn, created railway engineering corps. This new "cinematic" conception of logistics, or the "art of moving armies," led the Prussian general Von Moltke, creator during the Austro-Prussian War of the first "bureau of communication lines" (1866), to declare that he "preferred the building of railroads to that of fortifications."[33]

In contrast to the star-shaped system of the French railways, centered around the capital, the model that prevailed in imperial Germany consisted not only of tracks radiating in all directions from Berlin but also of a system of concentric lines circling the empire, so as to maintain communication between big cities in case tracks were destroyed in wartime.

This circular construction developed slowly. In 1834, the economist Friedrich List, promoter of the *Zollverein*, the union of customs and economies of the German states created at the initiative of Prussia, declared that "the railway system and the customs union are Siamese twins." After the political unification of Germany (1870), Chancellor Bismarck, on the advice of Von Moltke, sparked a movement to unify the networks, breaking down regional particularisms by nationalizing the major lines and organizing them according to military principles, an idea Von Moltke had been advocating since 1840, with his first strategic analyses of rail.

The Franco-Prussian War (1870) provided a practical demonstration of German superiority in organizing rapid movement of massive armies. The Boer Wars (1899–1902) and the Russo-Japanese War (1904–5) prefigured the strategic methods of World War I.

The Chronometer

The "rail model"—as it was dubbed by Paul Virilio—is above all a model of the administration of time. Witness this statement of M. Audibert, a

French polytechnician in charge of rail exploitation in those years of network rationalization: "If we all succeed, over the whole extent of the network, in respecting time to the second, we will have given humanity the most effective instrument for the construction of a new world."[34]

This mystique of controlling time, allied with the mystique of heavy industry, was also shared by the American engineer Frederick Winslow Taylor (1856–1915) who in the 1880s began to wage his campaign against workers' "loafing"and to apply his "scientific system" of labor organization in large steelworks. It is intriguing to note that in 1880 Caribbean writer Paul Lafargue, the son-in-law of Karl Marx, published *Le droit à la paresse* (The right to be lazy).[35] This apologia for pleasure, a mordantly ironic plea in favor of "the right to leisure," was to prove the most often translated text, after *The Communist Manifesto*, in socialist literature before World War I, appearing in languages from Russian to Yiddish.

In dissecting the movements of "the human animal" while perfecting new machine tools, Taylor broke all activity down into elementary, automatic operations. His idea was to induce workers to surpass the supposedly normal output by penalizing them if they did not attain it, and richly rewarding them if they surpassed it. This was Taylor's principle of differential wage rates. The use of the stopwatch made it possible to determine "scientifically" the base time for the manufacture of a piece. By reducing the number of people performing a given task and by substantially increasing the number of supervisors or inspectors, time-keepers or rhythm managers, accountants, and other managers of the time of workers subjected to the system—route clerks, instruction cards clerks, cost and time clerks, gang bosses, speed bosses, inspectors, repair bosses, the shop disciplinarian—the inventor of the first managerial doctrine was also proposing a scheme of internal communication. As Taylor himself wrote, "Dealing with every workman as a separate individual in this way involved the building of a labor office for the superintendent and clerks who were in charge of this section of the work. In this office, every laborer's work was planned out well in advance, and the workmen were all moved from place to place by the clerks with elaborate diagrams or maps of the yard before them, very much as chessmen are moved on a chessboard, a telephone and messenger system having been installed for this purpose."[36]

These ideas developed in a context where the question of labor organization gave rise, among its theoreticians, to the most daring projects regarding possible uses of new tools of communication. Thus, in his *Motion Study* (1911), the American Frank Bunker Gilbreth, a specialist in the study of micromovements and a pioneer of the rationalization of bricklaying, advised "intelligent bosses" to play music on a phonograph on their shop floors to fight boredom and monotony.

The concept of the separation of "tasks" and of the worker as a "human motor" would be ridiculed by the French anarcho-syndicalist Emile Pouget (1860–1931) in his work *L'organisation du surmenage: Le système Taylor* (The organization of overwork: The Taylor system), published in 1914 on the occasion of a workers' strike in the Renault auto factory against the American engineer's system.[37] These demands were blown away like straw in the wind by the imperative to mobilize production for war ends. The paradox was that in the United States one of the bitterest critiques of Taylor's disciplinary system of work was to come from rear admiral John Edwards, general inspector of machines in U. S. Navy warships. As quoted by Pouget, Edwards declared: "Taylor carries the system to such a degree that it becomes more of an obstacle than an aid in the efficient management and development of industry. It discourages mechanics, who do not feel motivated to show initiative by inventing improvements to remedy the faults inherent in construction."[38]

In the wake of the introduction of the stopwatch as an instrument to measure workers' motions, there also appeared an instrument for checking on their presence or absence. This was the time clock, whose true function was well described by an advertisement published in the *Almanach Didot-Bottin* in 1901: "Devices for controlling workers, patented in Germany and abroad. Latest novelty! New principle! The fastest, surest, simplest way of controlling the comings and goings of workers, while remaining invisible to them."

Numbers, Targets, Fingerprints

The need to manage large numbers engendered the need to process information. It was during the census of 1890 that the U. S. federal government first used a machine with perforated cards, invented ten years earlier by the American statistician Hermann Hollerith (1860–1929). It was based on the same principle as that perfected by Joseph-Marie Jacquard for his industrial loom. This first machine for processing information would be industrialized after 1896 by the Hollerith Tabulating Machines Corporation, an ancestor of IBM (International Business Machines), founded in 1924.

In the field of the media, this preoccupation with measurement and calculation was still in an embryonic phase. The modern American advertising agency was born in the early 1840s in Philadelphia; Volney B. Palmer is generally credited with being the father of the first agency. But it was not until 1865 that the concept of "target" was deployed. It originated with the J. Walter Thompson agency, which continued to be in the forefront of the refinement of approaches to the consumer in the following century. The first reflections on the "target" were made in the 1870s on the occasion of the first advertisements in the recently founded women's magazines (*Godey's Ladies' Book* and *Peterson's Magazine*, published in Philadelphia, the magazine center of America in this period), which were also the first periodicals that aimed to build a mass public. It was through women that access could be gained to the whole family targeted by the advertising message.[39] Not until the 1920s, however, would there emerge, under the aegis of Fordism, a "rationalization" of the targeting of consumers. In the meantime, in 1899, J. Walter Thompson was already installing a London office, the foundation stone of a whole network of branch offices that by the end of the 1920s was becoming international, following in the footsteps of General Motors, its first multinational client.

It should be noted that in the two last decades of the nineteenth century measurement and the need for classification obsessed those in charge of both the judicial and the penitentiary systems. In 1885, Rome saw the first congress of criminal anthropology, or "the science of the study of the delinquent," which signaled the point of departure of numerous national associations in this field and international links among them.

This congress, opened by Professor Cesare Lombroso, a doctor by profession and the author of *L'uomo delinquente* (Criminal man, 1876), had been preceded by the first international penitentiary congress, organized in London in 1882; it would be followed by the first congress of the International Union of Penal Law, to take place in Brussels in 1889, the year of the second congress of criminal anthropology, timed to coincide with the Universal Exposition in Paris. This new science, inspired by positivist philosophy, was eminently operational. It defined and classified the delinquent as an abnormal and dangerous individual, fundamentally psychopathic, "mentally ill." It was at this time that studies were made of anarchists, seen as "antisocial subjects" par excellence. This was also the time when the notion of "race" as a criterion for evaluating the "dangerousness" or the "intelligence" of the "foreign individual" revealed its profound ambiguity.

Anthropometry became a police tool for determining individual identity. A method of identifying delinquents known as *Bertillonage* (invented by the French anthropometrist Alphonse Bertillon, 1853–1914) was officially adopted in France in 1890. Here began the judicial identity photo, with shots face-on and in profile, combined with bodily measurements. Since 1871, date of the insurrection of the Paris Commune and its brutal suppression, state power had not ceased to display its concern for the use of photography for police purposes. In the 1890s, in the same perspective of clinical criminology, the Argentine police officer Juan Vucetich invented a fingerprinting system of criminal identification that rivaled Bertillon's method. We should not be surprised that Argentina manifested such concern in this domain. Like other countries in the Americas with a large influx of immigrants, Argentina's public authorities tried to better control the entry of "common delinquents" fleeing justice in their country of origin, and, above all, the entry of "elements arriving from Europe contaminated by dangerous and corrupting ideas." In their international congresses, Latin American specialists were not content with merely comparing the respective merits of Bertillon and Vucetich; they went so far as to propose the creation of "intercontinental offices of identification" to better control the immigrant influxes. The anthropometrist Sir Francis Galton (1822–1911), best known as the pioneer of eugenics, published his *Method of Indexing Finger Marks* in London in 1891, followed by *Finger-Prints* the next year.

Metaphors of Progress

In the second half of the nineteenth century, new actors and new forms of organization made their entrance onto the international scene. As we have seen, the first international organization of the modern age was created in 1865 to regulate the use of the electric telegraph.

After 1870, administrative unions and agreements between states multiplied. While only 20 were concluded in the 1870–80 period, there were 31 between 1880 and 1890, 61 between 1890 and 1900, and 108 between 1900 and 1904.[40] These unions and agreements concerned many different domains: the Universal Postal Union; the Red Cross (1874); an agreement signed by 25 nations on "universal time," fixed with reference to the Greenwich Meridian (1884); the Private International Law Agreement, concerned mainly with defining the forms of legal aid in interstate relations (1896); and others. This new type of organization inaugurated the new age of legal and technical internationalization. Codification and nomenclature were the order of the day.

In the appearance of these new spaces of international mediation, communications played a large role; it was tied up symbolically with the "bringing together of peoples." Throughout these decades, communications shared this mythology of "general concord" with other new forms of international relations: the universal expositions. They supported each other in a synergetic movement.

In 1851, the first universal exposition—the Great Exhibition of the Works of Industry of All Nations—took place at the Crystal Palace in

London. This was the occasion of the first telegraphic link between England and France. In 1876, the Philadelphia Exposition unveiled Alexander Graham Bell's telephone. In 1893, the World's Columbian Exposition of Chicago opened with the inauguration of the first telephone link between Chicago and New York. These are examples of how communications inventions were among the first ways of materializing—and also idealizing—notions of progress, civilization, the universal and universalism. These expositions were made to "show the degree of civilization and progress the various nations have attained," as Gérault put it in 1902.[41] In the international fairs of the Middle Ages (Beaucaire, Frankfurt, Leipzig, Lyon, Nijni-Novgorod), the exhibited objects were destined for immediate sale. The universal expositions, on the other hand, exhibited the machines used to make these objects. "They were organized to give an idea of the development of industry, trade and the arts in different countries," wrote Gérault.[42]

Self-promoted as "peaceful gatherings of progress," they did not escape the permanent tension characterizing the world at the time between the desire for general harmony and the impulse toward war. At the 1851 exposition, in the section occupied by the *Zollverein*, the electric telegraphs of the Siemens firm were placed alongside Krupp cannons. And 20 years later, at the Philadelphia exposition celebrating the centennial of U. S. independence, an official French report did not hesitate to stress, again with respect to the German section:

> One feels dominated by a painful feeling when, walking around the exposition, one finds nothing but entire German regiments, the emperor, the Crown Prince, Bismarck, Moltke and Roon, in porcelain, bisque, bronze, zinc, terra cotta, china—painted, embroidered, sculpted, printed, woven, etc. . . . In the machine section, seven-eighths of the space is occupied by the Krupp cannons, *killing machines* as the Americans call them, which seem a brutal menace among the peaceful exhibits of other nations.[43]

And the novelist Emile Zola could not find words strong enough to lambast the 1867 Paris Exposition, which he described as an "imperial festival," an "extravanganza of lies," "stuffed with majesties and highnesses" where "the crowds thronging the exhibitions made a popular success of the Krupp cannons, huge and dark, exhibited by Germany."[44]

After London's 1851 exposition, the largest of these fairs took place in Paris (1855, 1867, 1878, 1889, 1900), London (1862), Vienna (1873), Philadelphia (1876), and Chicago (1893), each trying to outdo the other in attendance records. The 14,000 exhibitors at the Crystal Palace attracted 6 million visitors; the 83,000 exhibitors at the 1900 Paris Exposition drew nearly 51 million. The Paris Exposition of 1889, which celebrated the centennial of the Revolution, covered 100 hectares; that of Chicago, celebrating the fourth centenary of Christopher Columbus's expedition, was five times as large but drew only 27 million visitors.

The gigantic proportions of the World's Columbian Exposition in Chicago reflected the large political stakes. Reinterpreting Columbus's feat, the young American nation sought to swing the pendulum of the international order toward Pan-Americanism. In the words of anthropologist M. R. Trouillot, the Pan-American strategy was intended "in part to block the British (whose investments in South America exceeded those of the United States), the French (perceived as a major threat until the 1889 collapse of their Panama Canal project), and, to a lesser extent, the Germans and Italians."[45]

In 1889, Secretary of State James Blaine invited Latin American countries to the First International Conference in Washington. Its purpose was to promote continental peace, to sign agreements on customs and trade, and to formulate a plan of arbitration for the settlement of disputes. The results of the meeting did not match Washington's hopes. Its most important accomplishment was the creation of the Commercial Bureau of American Republics (later changed at its fourth conference in Buenos Aires in 1910 to the Pan American Union).

The last quarter of the nineteenth century saw the proliferation of the form of communication known as the international exposition. Numerous cities organized such events, without attaining the scope of the great expositions already mentioned. Between 1879 and 1889, there were no less than 25: Sidney (1879), Melbourne (1880), Bombay (1887), Barcelona (1886), Edinburgh (1886), and many more. To this long list must be added the first great specialized exposition: the International Exposition of Electricity in Paris (1881). In the preceding decade, some Latin American capitals had also organized expositions: Lima (1872), Bogotá (1872), and Santiago de Chile (1875). The first international exposition on African soil was organized in the Cape of Good Hope, at Grahamstown, in 1898.

The cosmopolitan rhetoric of universal fraternity and the people's fair scarcely conceals the fact that the universal exposition was a place of rival nationalisms and the production of a public discourse—political and scientific—that consecrated the notion of "Western civilization" as the beacon of progress for other peoples. Professor Michel Chevalier, reporting on the Paris Exposition of 1867, wrote enthusiastically:

> Civilization has displaced its original home. . . . After India and
> Egypt, Chaldea and Greece; after Rome, the great triad of modern

Europe: France, England, Germany. It is in these latter regions that the forces of the human spirit have most developed, and that morality, science, and industry have taken on a superiority to everything that came before. . . . In the regions that have been relegated to an inferior rank, the genius of Europe opens territories through perfected communications.[46]

He added: "I use the term Western civilization instead of European civilization because of the United States, which one cannot consider separate from it, since the people of that country practice the same arts, follow the same methods, and generally live on the basis of the same religious, moral, social, political, and scientific ideas."[47]

After 1874, Friedrich Nietzsche discerned beneath the grand celebrations of universalisms the morbidity of the European expansionist instinct:

The Roman of the Empire ceased to be a Roman through the contemplation of the world that lay at his feet; he lost himself in the crowd of foreigners that streamed into Rome, and degenerated amid the cosmopolitan carnival of arts, worships and moralities. It is the same with the modern man, who is continually having a world-panorama unrolled before his eyes by his historical artists. He is turned into a restless, dilettante spectator, and arrives at a condition when even great wars and revolutions cannot affect him beyond the moment. The war is hardly at an end, and it is already converted into thousands of copies of printed matter, and will be soon served up as the latest means of tickling the jaded palates of the historical gourmets.[48]

A quarter of a century later, John Atkinson Hobson, one of the first analysts of modern imperialism, denounced the universal expositions even more firmly:

It is obvious that the spectatorial lust of Jingoism is a most serious factor in Imperialism. The dramatic falsification both of war and of the whole policy of imperial expansion required to feed this popular passion forms no small portion of the art of the real organizers of imperialist exploits.[49]

What is beyond dispute is that the new scenario of international exchanges, to which the ephemeral shop windows of the universal expositions bore witness, had since the end of the nineteenth century profoundly changed people's representations of the world and their ways of experiencing the relation between national and international. The intricate network of communications, banking and insurance services, the great trans-Atlantic migrations, the expansion of multilateral trade through the international division of labor, together brought the ideas of "interdependence" and a "system of interdependence." As historian Douglas McKie wrote:

> The tacit assumptions on which the system of interdependence was based concerned both the relationship of countries to each other and of governments to their peoples. The segregation of economics and politics, reflected in the limited extent to which politicians interfered with international economic specialization, depended largely upon the social framework and social pressures inside their countries. . . . The conception of a "legal" or "moral" order behind this intricate pattern of interdependence was challenged most articulately before 1914 at moments of business recession, as in the years 1907–1908 when there were sharp downswings in the volume of economic activity and employment.[50]

The search for this new international order took place in a world in which the structure of relations among states was still fundamentally determined by the fear-driven policies of the great powers. The tension between the logics of negotiation and those of security/insecurity was too tangible to render credible the first efforts to construct a system for regulating international relations. Examples are the first Hague Peace Conference of 1899, whose purpose was to restrict competition in the accumulation of armaments, and the Conference of Algeciras in 1906 that aimed to settle colonial disputes. It was only much later that the founding of the permanent court of arbitration, approved at the 1899 Hague Conference, revealed its pioneering character. In the short term, the arms race was to continue and the two conventions relating to the rules of war signed during this first world peace conference would carry little weight when confronted with the violence of World War I.

How did theory account, during the nineteenth century and the beginning of the twentieth, for the rise in the power of technical networks of communication and the setting up of mechanisms of mass opinion? That is the question broached in the next chapter.

Chapter 2

The Age of Multitudes

The Promise of a New Moral World

"We can understand nothing rightly unless we perceive the manner in which the revolution in communication has made a new world for us."[1] This would be a banal statement had it not been made in 1901, a dozen years before the birth of Marshall McLuhan, when few imagined what future the technical networks reserved for the industrial society still in construction. Its author is Charles Horton Cooley (1864–1929), who is considered one of the founders of modern American sociology.

Communication is the mechanism by which society organizes itself, Cooley believed, and the mechanism thanks to which human relations exist and develop. It is a double mechanism: there is physical or material communication, concerned with transport, participating in the physical organization of society; and psychological communication, a veritable agent of social organization, including symbols and all the apparatuses that make possible their conservation and transmission. The field of communication has as much to do with facial expressions, attitudes, gestures, tones of voice, and words and styles of writing as it does with printing, railways, telegraph, and telephone—in short, it is everything that may result from the mastery of space and time.[2] This "new world of communication" entails a fundamental change in mentality: in Cooley's terms, *enlargement* of mental perspective, and *mental animation*, the product of frequent exposure to novelty. This is because the use of new means of communication fulfills four functions: expressiveness, or the range of ideas and feelings these means are equipped to convey; the permanence of record, or the overcoming of time; swiftness, or the overcoming of space; diffusion, or the access to all classes of men. But this innovation

31

has a price. New dangers lie in wait for individuals subject to the increase in the "intensity of life" produced by the circulation of ideas and images: superficiality and the strain caused by difficulties in understanding and assimilating all that is new. The risk therefore exists of a rupture of the personality: depression, suicide, or madness.

Cooley did not neglect the observation of what he called "primary groups," such as the family and youth, or neighborhood groups, where people develop face-to-face relations based on association and cooperation. These relations remain essential because within them is formed, through the process of communication, "human nature," "sentiments . . . such as love, resentment, ambition, vanity, hero worship, and the feeling of social right and wrong"—in short, all that makes possible "sympathy," that is, an "imaginative contact with the mind[s] of others," an entering into and sharing of the mind of someone else.[3] Cooley was disturbed at the vulnerability of primary groups to the effects of mounting urbanization. But he was also able to assess the capacity of neighborhood and family to adapt and redefine themselves in the face of the new conditions of modern anonymity.

Resolutely optimistic despite the tensions and ambiguities that characterize the means of modern communication, Cooley's reforming spirit saw in the "new communications" an instrument for attaining an era of moral progress. With the advent at the world level of a common sentiment of belonging to humanity and the infinite growth of justice, communication would destroy the old social order and build a new one in which people would increase their "sympathetic" contacts.

We must remember the origin of this conceptual framework. Before developing his first hypothesis on the meaning of the new mode of social organization made possible by communication, Cooley had made two studies, one for a railroad regulation commission recently created under the Interstate Commerce Act, the other for the Federal Census Bureau. The first bore on the search for ways by which railways might cut down accidents, the second on the social significance of street railways. His first book, published in 1894, took as its title *The Theory of Transportation*.[4] In fact it was his Ph.D. thesis, defended before a committee at the University of Michigan, which counted among its illustrious professors at the time two other leading figures on the American academic scene: the social philosopher John Dewey and the social psychologist George H. Mead. In the history of American social science, all three were representatives of the tradition of critical reform, sharing the hope of seeing communication serve the renewal of democracy.

This first attempt to map the sociological territory was of interest for at least two reasons. First, when Cooley formulated the distinction between physical and psychic communication, he brought to light the difficulty of articulating the various levels of analysis. His trial-and-error procedure augured the split between the two poles that was to recur time and again in the history of theories of communication in the technical age. This is a split that Marx himself tried to resolve at the level of his global theory of society by proposing the dichotomy between economic infrastructure and ideological superstructure, producing a vision of a pyramid cut into sections, or instances; its interpretation was to be a permanent source of argument among his followers, and it long deferred sufficient consideration in Marxist analyses of culture and symbolic practices. Sometimes taking refuge in the symbolic, sometimes in the material, placing sometimes one, sometimes the other in the foreground and rarely treating them at the same time, this tendency of communication theory was to recur and even intensify as two conceptions grew apart: one that limited communication to the exchange among individual psychologies and the other that refused to think of the individual subject outside of his or her social relations. The two were rarely combined. But for Cooley, sociology was not worth its name unless it tried to define itself as the study of "personal relations, at the same time in their primary aspect and their secondary, as groups, institutions and processes."

The second interesting aspect of Cooley's work is that it allows us to perceive how the study of communications, from its first stammerings, was invested with the hopes of a social revolution. In Cooley we see one of the first theoretical manifestations of this communicational millenarianism. Endowed with a redemptive function, communication offers the promise of a new communion, or a new community. This should hardly surprise us, since a work published in 1852 under the title *The Silent Revolution* had already predicted a new social harmony to be attained thanks to "a perfect network of electric filaments."[5] This millenarianism would later underpin a number of discourses on the new "informational agoras." Like the tendency to bipolar analysis mentioned above, it too is therefore constitutive of the history of communication, its theory, its doctrines, and their uses.

Universal Association

In the emergence during the nineteenth century of a belief in the salutary determinism of techniques of communication, the disciples of the philosopher and economist Claude Henri Saint-Simon (1760–1825), precursor of positivist social science, played an essential role. The doctrine of Saint-Simon, who deeply influenced Auguste Comte, is even more crucial in that from among his disciples emerged both inventors of communicational utopias and great builders of the means of communication. In France they were the pioneers of railways and trans-Atlantic steamship lines, and were closely linked to the great enterprises of the Suez and Panama canals, which revolutionized world geography. They were undoubtedly the first to formalize the notion of "network," as a way of conveying the rupture brought about by the new techniques of trade and circulation of goods, messages, and people.

From the very beginnings of the railway, the Saint-Simonians endeavored to calculate the moral and social consequences of the arrival of this new network. From an international point of view, the work of the Saint-Simonian Michel Chevalier (1806–79) entitled *Le système de la Méditerranée* (The system of the Mediterranean), published in 1832, remains the most daring. This future president of the international jury of the Universal Exposition of 1867 was convinced that "the railway is the symbol of universal association." Seeing it as a means to reconcile the West and the East around the Mediterranean, he designed an imaginary network by which this new trait of universal union could be transmitted. Tracing his project of a "European Confederation" by means of rail, the author takes us successively to Spain, France, Italy, Germany, European Turkey, Russia, Asia, and Africa, and with the seven-league boots of networks, he jumps the Bosporus to the Persian Gulf, from the Persian Gulf to the Caspian Sea, from the Elephantine to Alexandria, and so forth. As for the impact the railway could have on Russia, he wrote: "If there exists one country where the railways should have a civilizing influence, it is Russia. Everything slumbers among the inhabitants of this country, who die after having vegetated rather than lived, without going beyond the sight of their ancestors' hovels, resembling mollusks whose shells are fixed to a rock. In the political realm, the most effective means to awaken them from their somnolence would be to place near them examples of an extraordinary movement, to excite them with the spectacle of prodigious velocity, and invite them to follow the current that will flow to their door."[6]

But the destiny of most of the disciples of Saint-Simon also shows how short is the distance separating certain communicational utopias from pragmatism as soon as their creators gain power. Converted to businessmen or governmental advisors, the Saint-Simonians grew further and further away from their youthful ideal of socialism, and turned into ardent defenders of free trade and industrialism. The apostle of "universal association," Michel Chevalier, became a professor of political economy at the University of Paris, and as advisor to Emperor Napoléon III in the 1860s, he invented "Pan-Latinism." This doctrine, conceived as a response by the imperial state to Washington's Pan-Americanism, legitimated the intervention of French troops in Mexico (1862–67) to defend the emperor Ferdinand-Joseph Maximilian (who would be executed in 1867), placed on the throne by Napoléon III. As in Napoléon Bonaparte's intervention in Egypt, the expeditionary force was backed up with a corps of scientists and engineers—"our barbaric civilizers," as they were called in 1862 by the president of the Mexican Republic, Benito Juarez.

It remains true that, shorn of its early utopian aspirations, Saint-Simonianism left its imprint on the construction of social science. Social science, in this perspective, was to be founded on the same positive and objective principles as physical science, and was to apply the same methods: it would be both a "science of observation" and a "science of organization." In publishing in 1821 the first volume of *Le système industriel* (The industrial system), Saint-Simon had laid the foundation of his "new science of man" by means of the concept of "system." The concept was meant to reflect the quest for explanatory models of the social totality in an instrumental perspective: How should exploitation of the planet by industry be organized? How should we think and act in order to deploy "the law of progress"—that "fundamental law of historical development"? Saint-Simon thought that only a social doctrine that proposed a general concept of society, organized it and made it intelligible, could move society from a "critical state" to an "organic state." In the critical state, all unity of thought and action ceases among men, and society is no more than an agglomeration of isolated and competing individuals and personal interests. In the organic state, all activities are ordered around a definition of the goal of social action: society becomes one.

The Social Organism

What is the nature of the nascent industrial society? What is the significance of social change? This enigma, which the American Charles H. Cooley tried to resolve on the basis of "communication," the Frenchman Auguste Comte (1798–1857) had already tried to answer in the 1830s, even before the great epoch of industrialization, when he theorized the "social dynamic" and development of industrial society.

At the foundation of his thought was the idea that the social organism only succeeds in finding harmony and stability via the division of labor. The more a society grows, the more its constitutive parts differentiate. This division of functions, fulfilled by more and more differentiated but interconnected organs, united in their purpose but diverse in the means they employ, is the golden rule of stability in a society—as well as a source of disorganization. In a society imagined as a "system"—the whole greater than the sum of its parts—where social equilibrium is the result of the individuality of work and of cooperative effort, the danger lies in an excessive division of tasks, an exaggerated specialization.[7] These are the hypotheses, inscribed in an organic conception of society, which would furnish the recurrent frame of reference in the debate on industrial society as a "mass society," and more particularly, on the question of whether mass communications has the ability to assure the bonds between individuals that allow for the maintenance of an integrated and stable system of social control.

These conceptions—of the laws of evolution and of the social order as an organism—would be extended by the British thinker Herbert Spencer (1820–1903), who would carry to their logical extremes notions such as growth, structures, functions, or systems of organs, and further develop the complex analogy between society and organism. Unlike Comte, who saw a degree of planning as necessary for change, Spencer was a militant of laissez-faire,[8] in this closely associating himself with the search for legitimation of a bourgeoisie then rising to the position of command in the industrializing process of his country. Spencer, in a context dominated by the evolutionary theses of Charles Darwin, displayed in its full force the tendency to organicize or "biologize" the social. Under the auspices of positivistic tradition in sociology, there began to emerge, starting at the end of the nineteenth century, the idea of communication as the regulatory principle counteracting the disequilibria of the social order.[9] This conceptual matrix would later reach its high point in the functionalist sociology of mass communications; the "religion of progress," so dear to the first positivists, would in the following century metamorphose, by various stages, into the "religion of communications."

The influence of Darwin and Spencer made itself felt in the last years of the nineteenth century, particularly in two domains that one way or another are concerned with thinking about information and communication.

The first is that of the sociology of leisure, associated with Thorstein Veblen and his masterwork *Theory of the Leisure Class*, published in 1899.[10] In this book, Veblen, who was also influenced by American pragmatism, develops his theory of conspicuous consumption: the meticulous search for quality in food, drink, home, services, ornaments, clothing, and amusements is not so much to satisfy real needs as to elevate or express social *status*, to maintain social prestige. Functionalist sociology of the twentieth century was to see in this theory a decisive contribution to the concept of "latent function": a function that, in contrast to the avowed "manifest function," is neither willed nor generally recognized in its social and psychological consequences.[11] The critical school, on the other hand, interpreted Veblen's thesis as the first paradoxical expression of a sociology of culture combining positivism and a demystifying approach. Veblen, who was close to the Rousseauist ideal of the primitive, denounced the barbarism of modern times—"barbaric civilization," the barbarous character of a culture given over to ostentation and to advertising. At the same time, this man who despised the culture he was analyzing in its most banal phenomena—from the sports referee to lawns to nineteenth-century interior decoration—suggested no recourse for individuals other than an attitude of acceptance of what existed. His central concept was "adaptation," the adjustment to the new Darwinian world of natural selection in production and industrial technology. Science, conceived as the universal application of the principle of causality in opposition to archaic, animistic forms of thought, is the instrument of progress. The new forms of "life" to which one ought to adapt all refer back untiringly, in Veblen's thought, to the sphere of economic consumption.[12]

Life Space

The voguish ideas launched by the science of the social organism also marked the first analyses of modern geopolitics, which started to theorize the new spatial dimension of international exchanges. Its precursor was the German Friedrich Ratzel, author of a voluminous treatise on "political geography" published in 1897.[13] His research threw up new spatial concepts, such as spatial representation (*Raumvorstellung*), life space (*Lebensraum*), concentration (*Zusammenfessung*), frontiers (*Grenzen*), and world power (*Weltmacht*). With his notion of "circulation" in the sense of transmission of information, he was one of the few people to perceive the structuring role of the new networks of telecommunications, to which he devoted a chapter in his treatise. "Space (*der Raum*) is power," postulated Ratzel, whose thought would later be echoed in the organicist thinking of the Swedish geopolitical analyst Rudolph Kjellen in a 1916 work entitled *Staten som Lifsform* (The state as life-form). Both were trying to lay the foundations of strategic policy in the analysis of the geography of nations. "Position" (*die Lage*) was, after space, the second key concept in their approach. Space or territory was a matter for the state, an imperfect organism whose development is governed by laws regulating the *functions of gestation: birth, growth*, and *strengthening*. In order to grow, the state needs the necessary space (*Lebensraum*).

In this conception of the state as a living being, "circulation" is a capital factor: it "vitalizes" territory, channels pressures, orients defensive reactions, gives a concrete meaning to breadth, form, and situation. It links together internal and external political spaces.

This first exploration of the concept of space coincides with what the French geographer Roger Brunet has identified as the "spatialist ideology," which he sees as a correlate to biologism. Both would prove fertile sources for the legitimation of expansionism, as is demonstrated by the later use history made of the notion of "life space" or, more subtly, "natural borders." Vital space became one form of the animal territorial law, justifying war, conquests, and encroachments; "natural" space meant the necessary mastery over resources and supplies, which legitimated efforts to control the spaces on which the state "depends," whether they be oil wells, copper mines, or uranium deposits.[14]

The last decade of the nineteenth century thus saw the formation of modern imperial doctrine. It is useful to recall that in the elaboration of his political geography, Ratzel had himself been influenced by a work by U.S. Navy Captain Alfred Thayer Mahan (1840–1914), *The Influence of Sea Power upon History* (1890). Written in the years when a naval armaments buildup and a "big Navy" policy reflected America's rise to industrial power, this book had been preceded by two other works that also placed U. S. expansion on the agenda as a necessity: *Our Country* (1885) by Josiah Strong and *Manifest Destiny* (1885) by John Fiske. Mahan did not beat about the bush: "Arrested on the South by the rights of a race wholly alien to us, and on the North by a body of states of like traditions to ours, whose freedom to choose their own affiliation we respect, we have come to the sea. In our infancy we bordered upon the Atlantic only, our youth carried our boundary to the Gulf of Mexico. Today our maturity sees us upon the Pacific. Have we no right or no call to progress in any direction?"[15]

The response was not long in coming, furnished by the events of 1898: the arrival of U.S. naval vessels in Philippine waters, the landing of Marines in Cuba, and the annexation of Puerto Rico and Guam. The strategic idea of Mahan was the following: "The due use and control of the sea is but one link in the chain of exchange by which wealth accumulates; but it is the central link, which lays under contributions other nations for the benefit of the one holding it, and which history seems to assert, most surely of all gathers itself riches."[16]

In 1902, Captain Mahan was elected president of the American Historical Association, an honor never before granted to a man in uniform.

This geography of world power developed by Mahan, Ratzel, and Kjellen contrasted at the time with that of another pioneering approach to geography represented by a Frenchman, Elisée Reclus (1830–1905), who had been banished from France in 1872 for having taken part in the

Commune. Between 1875 and 1894 there appeared 19 volumes of his *Nouvelle géographie universelle* (New universal geography). In contrast to the great majority of his contemporaries in the "beacon societies of progress and civilization," where geographers played a not insignificant role in the preparation of European expansion, Reclus had, in producing his atlas, initiated more egalitarian relations with scientific circles in countries that would much later be called "developing." He relied upon a vast network of local researchers, including several Latin Americans, and incorporated their knowledge into a new vision of the planet that recognized its profound inequalities. This was particularly true for the three volumes devoted to Latin America, where the young geographer Reclus had begun his career by producing a study of the region of Santa Marta, Colombia.

At the end of his *New Universal Geography*, the French geographer wrote in 1894:

> I wanted to make my story live, showing for each country the traits that characterize it, indicating for each group of humanity the genius peculiar to it. I would say that everywhere I feel at home, in my country, with my brother men. I do not believe myself carried away by a sentiment other than that of sympathy and respect for the inhabitants of a common nation. On this planet that turns so quickly in space, a grain of sand amid immensity, is it worth it to hate each other? But in placing myself within this view of human solidarity, it seems that my work is not complete.[17]

In fact, Reclus was faithful to the grand themes and beliefs of the anarchist movement of which he was a theoretician, along with the Russian Prince Peter Kropotkin (1842–1921), his companion in militancy. Human solidarity and universal fraternity, which Kropotkin called "the natural principle of mutual aid," was according to him stronger than the Darwinian principle of struggle for existence. While Reclus, in his books, was attentive to new networks of communication, Kropotkin went much further in his vision of the impact of these technologies: he saw in electricity a means of reconstituting lost community. In this he prefigured a current of thought centered on the theme of technical civilization.

Mass against Community?

At the turn of the century, other founding fathers of social science developed their own perspectives on the specificity of the changes taking place, and left a heritage of key concepts that would later help to structure the sociology of mass communications.

For example, the German Ferdinand Tönnies advanced the debate on mass society by proposing that we distinguish between two theoretical constructions, one representing the society that Europe was in the process of leaving, and the other, that she was entering: respectively *Gemeinschaft* and *Gesellschaft*, or "community" and "society." The former is by nature affective and existential, characterized by informal social relations. The latter is a distant relative of that complexity envisioned by Comte in his notion of division of labor. It is rational by nature, constructed around a contract accepted voluntarily by its subjects, each party agreeing to certain obligations and accepting sanctions if the clauses of the contract are not observed. The individual must face an impersonal, anonymous, and competitive system.[18] For Tönnies, these two poles were not mutually exclusive; mass society, urbanized and industrialized, did not imply the end of "primary groups," but their redefinition. Rather than representing structures and institutions, the two poles expressed two dimensions of social action. In any case, Tönnies avoided attributing positive value to one or the other.

This heuristic tool developed by Tönnies gave rise to several contradictory interpretations, some of which went to the extreme of overlooking what had inspired it. This realization would come later when the sociology of mass communications was confronted, after World War II, with intercultural relations between so-called developed or modern societies and those referred to as traditional.

The German sociologist, in the manner of other contemporaries like the American Charles H. Cooley, or his compatriots Georg Simmel and Max Weber, or the Frenchman Emile Durkheim, insisted on the ambivalence of change and on the contradictory character of the new world, which was seen to be just as capable of bringing more freedom and pluralism as of engendering more dehumanization, anonymity, and fragmentation, as well as more state oppression and bureaucracy. They all thought that people should confront the growing process of social differentiation and the establishment of new social relations in modern cosmopolitan society. They believed that, even if a process of homogenization was under way, the "theory of zero level," as Cooley called it, had no basis whatsoever. In other words, they opposed the idea that pluralism and the search for equality necessarily brought about a leveling down, or the triumph of the "lowest common denominator."

Fear of Crowds

The idea of the degeneration of society did, however, have some defenders: the "psychopathologists," who far from being ambivalent, developed a binary vision in which the pole of "mass society" was afflicted with all sorts of defects.

Whereas a sociologist such as Simmel saw a complex tissue of multiple relations resulting from continual interaction among individuals, in which society and the social were conceived of as production and process, the philosopher and psychopathologist Gustave Le Bon, author of a key work, *Psychologie des foules* (*Psychology of crowds*), published in 1895, saw "the delirious crowd" and "mental contagion": mass society produced "automatons no longer guided by their will."[19] An implacable enemy of the principle of equality and champion of the national heritage against cosmopolitanism, Le Bon interpreted the rise of "mass society" and the "dangerous classes" (the intrusion of masses into the city) as a mortal threat to elites and property owners. If one did not want to resign oneself to being submerged by the tide of uncontrolled violence, it was urgent to channel it "like an engineer masters a torrent." While Simmel was interested in the perpetual process of social production—in "sociation," as he called it—and intervened directly in the debate on the "woman question" and the rise of the women's movement in imperial Germany, linking the question of modernity and that of femininity,[20] Le Bon could find nothing better to do than to rail against the sentimentality of crowds, which "one observes also in the case of beings belonging to inferior forms of evolution such as women, savages, and children."

Aristocratic fear of the rise in power of the common people, in its modern version, had a strong antecedent in the theories of the English pastor Thomas R. Malthus, author of the *Essay on the Principle of Population* (1798).[21] In this work, which originally was conceived as a reply to the socialist doctrine of William Godwin (just as Le Bon directed many of his writings against his socialist contemporaries), Malthus imputed to the "laws of nature" the misery and unhappiness of inferior classes of people, for whom he proposed, as the only way out of their state of poverty, a reduction in their rate of proliferation.[22] Malthus, who died in 1834, aligned himself clearly with the interests of protectionist landlords who— since the compromise of 1689, the fruit of the "Glorious Revolution"— governed Britain alongside the bourgeois minority. He did not perceive or refused to admit the irresistible rise of this new manufacturing bourgeoisie, which was settling into power after 1846 with the abrogation of the Corn Laws and the triumph of free trade. It was in this process precisely that Spencer rooted his sociology of the social organism. Our willfully summary correlation between theoretical views and class interests is important to understanding what happened later, after World War II, when the sociology of communication and the sociology of population lent each other support in their approaches to poverty. It is not insignificant that the father of structural-functionalist sociology in the United States, Talcott Parsons, saw Malthus as a precursor because he was the first to formulate a theory of the regulatory function of institutions and of social equilibrium.[23]

In the debate in the 1890s in which Le Bon took part, the very status of public opinion was at stake. France had only recently consecrated the exercise of freedom of the press, thus taking its first steps toward "mass democracy," with large-circulation papers, the appearance of new political organizations of social solidarity, the gestation of modern public opinion—and also a major test, the Dreyfus Affair. This political and legal controversy, over the condemnation and deportation of a French captain of Jewish origin who was wrongly accused of espionage after a campaign of anti-Semitic hysteria, divided the country into two camps. Defending Dreyfus, who became a symbol and a pretext, was the League of the Rights of Man, founded by republican intellectuals; its opponent was the League of the French Nation, which took the side of the army and the church. One of the climaxes of the long polemic was the publication of "J'accuse" by Emile Zola in the paper *L'Aurore* (1898). This controversy marked the emergence in France of the intellectual class as an actor in the public space—a group that in the following decades would in its majority occupy the left of the political spectrum.

It was in this context of political turmoil that the work of Gabriel Tarde, a precursor of modern social psychology, gained importance. Tarde was contested by Durkheim, who in the name of sociology's autonomy as a discipline criticized him for drifting into psychologism. Tarde developed his early hypotheses on the relation between the media and the formation of opinion in opposition to the conception of the mob as subject to "criminal suggestion."[24]

Tarde made his mark by developing the notion of "public[s] inclined to imitation." Imitation, he said, is the rule of sociability, but contestation also exists. Between the two poles emerges the possibility of "invention" and the spread of "new ideas." The rise of the press and sensationalistic reporting has enlarged the group of actors in the formation of public opinion. Contrary to what Le Bon argued, the era of masses assembled in crowds and infusing frenzy throughout the body social was already a thing of the past: the era of publics had arrived. Publics are the "social groups of the future." The crowd is a primitive phenomenon determined by "action in the eyes of others"; the public is a cultured and civilized phenomenon, determined by "consideration of the outlook of others." One cannot be part of more than one crowd at a time; on the other hand, one may belong to several publics, dispersed and fragmented multitudes, but "mentally united." There lies the difference separating intolerance from tolerance. While Tarde posed the problem of the conditions of existence of the new mass democracy, which he linked to the new means of mass communications, Le Bon had a more static view of the "collective" as the state of a crowd that is "impulsive, irritable, incapable of reason and critical thought, agitated by their overemotionalism."

The theory of Gabriel Tarde would not lead to a French school; the status of social psychology, countered by a hegemonic sociology, remained precarious in the absence of institutional support. On the other hand, it would help form the foundation of the study of attitudes and public opinion in the United States.[25] As for the theses of the anti-Dreyfusard Le Bon, they were to become prominent in military circles on the eve of World War I, particularly in the French military academy. His *Psychology of Crowds* would long remain bedside reading for warriors of all nationalities,[26] well after World War I. Psychopathology managed to be at the same time a reference point for Marshal Foch, General De Gaulle, and Adolf Hitler, who went so far as to plagiarize it in *Mein Kampf.*

Internationalization of Trade and Concepts

In the elaboration of internationally shared schemas of analysis, the universal expositions occupied an important place. Above and beyond the spirit of the imperial times that inspired them, they fulfilled a historical function of mediation.

Their role as places of international exchange manifested itself above all through various meetings, colloquia, lecture series, and conventions on the margins of these events. This was all the more true after 1878 when the organizers of the Paris exposition decided to institutionalize this type of gathering.

In the first place, the exposition was the occasion for searching in concert for legal norms and techniques felt to be worthy internationally. Thus the London exposition of 1851, by making apparent the difficulty in measuring and comparing products, gave rise to a convening in Brussels, under the presidency of the Belgian Adolphe Quételet, of the first international statistics congress. Its deliberations resulted in the creation in 1875 of the International Bureau of Weights and Measures, and, ten years later, in the foundation of the International Statistics Institute. At the Vienna Exposition in 1873, a congress on industrial copyright proposed the first international convention on patents. A congress on literary copyright was organized at the Paris exposition of 1878 under the presidency of Victor Hugo. Eight years later the International Union was created in Berne for the protection of works of literature and art; an agreement was signed by ten states. During the international exposition of electricity in 1881, the Ampère unit was adopted as the basis of a universal electrical language. In contrast to nonspecialized expositions, where most sovereign nations participated, only sixteen nations were represented there to lay the groundwork for the new electrical science and industry: Germany, the United States, Austria, Belgium, Denmark, Spain, France, Hungary, Italy, Japan, Norway, Holland, the United Kingdom, Russia, Sweden, and Switzerland.

The universal exposition was also a place of exchange for a broad variety of social organizations. In 1862, at the London exhibition, workers' delegations took part and discussed numerous questions relating to forms of association (cooperatives, trade unions). It was no coincidence that the address of the Paris workers' delegation would serve, two years later, as the headquarters for discussions about the statutes of the First International. In 1876, in Philadelphia, feminist organizations were allowed to take part. In 1889, in Paris, several leagues of universal peace and freedom were particularly active. In 1893, the World's Fair in Chicago included a Board of Lady Managers constituted on the basis of equality with the heads of other departments at the Fair. In addition to organizing their own pavilion, this board launched a women's congress.[27] Chicago also featured conventions on the international role of the press, on religious congresses, and more. An International Feminist Congress took place within the Paris exposition of 1900.[28]

Finally, the expositions were occasions for societies of experts and scientists to evaluate the state of their disciplines: geographers, economists, psychopathologists, doctors, chemists, architects, meteorologists, ethnologists, and others. Thus in 1889 Gabriel Tarde participated, with the Italian Cesare Lombroso, in the second international congress of criminal anthropology. In 1900, at the International School at the Paris exposition, the Scottish professor Patrick Geddes, another pioneer of thought on "technical civilization," gave a lecture called "Elements of Progress in the Exposition: Neotechnic Elements."

A number of scientists, thinkers, and researchers lent their direct help to the organization of these expositions. In 1855, the French socialist Pierre-Joseph Proudhon drafted a report in which he proposed the organization of a "perpetual exposition." The geographer Elisée Reclus was associated with the Paris exposition of 1900, for which he conceived a terrestrial globe in relief, 26 meters in diameter. In Chicago, the German Franz Boas, one of the pioneers of classical ethnology and a precursor of functionalism, was in charge of the anthropological exhibits.

New forms of circulation of knowledge, new synergies between experts and industrialists, new modes of interdisciplinarity, new types of relations between science and art, and industry and art, gradually appeared. It was an era, too, when magazines such as the *Gazette des Beaux-Arts* (1859) and *Scientific American* (1845) were building a readership for popularized science.

How did the participants in these congresses evaluate the universal expositions under whose umbrella they took place? Here is one writer's reaction to the 1889 exposition: "The results of the congresses from the viewpoint of work accomplished are very real in certain cases: one may arrive at an understanding on common endeavour, on the rules to follow in nomenclature, on steps to take and research to do. In other cases, the congresses provided precious information, which, along with that already possessed, made it possible to sum up the subject as a group, or contributed to the completion of an enquiry."[29]

But this new type of international contact was also a source of new forms of unequal exchange. In the name of modernity, the model of urban management developed by the architect Hausmann for Paris was exported to Buenos Aires, to Santiago de Chile, and to Rio de Janeiro, in a Brazil that chose in 1889, date of the overthrow of the emperor and the foundation of the republic, to inscribe on its national flag the motto of Auguste Comte's positivism: "Order and Progress."

European criminal anthropology exported its conception of the profile of delinquency and normality to all of Latin America. Cartesian models of teaching emigrated to Latin American shores, and the "bovaryste elites," as the Brazilian Celso Furtado called them, lived to the rhythm of the Parisian theatrical season. Meanwhile the "Orientalist" anthropologists, grouped likewise in an expert society at the end of the century, invited the "Orientals," in the course of a universal exposition, to see themselves through the mirror offered them by the civilized West. Indian and Pakistani researchers have since demonstrated the role played by expositions—in close collaboration with local schools of art and museums—in the degradation of Indian art and crafts. In the decade from 1882 to 1892 there were at least seven major exhibitions featuring Indian arts and crafts, with venues in Lahore, Calcutta, Delhi, London, Glasgow, Paris, and so on. These expositions were systematically organized to promote the trade in Indian art manufactures by advising or guiding artisans or supplying them with designs.[30] *The Journal of Indian Art* (1883) and later *The Journal of Indian Art and Industry* were created for this purpose; the main responsibility for the production of the latter fell to Rudyard Kipling, the director of the Mayo School of Art, one of the four institutions founded in India around 1872 by the British authorities to train natives in "industrial art."

The Exposition of 1900 marks the peak of the ascending curve of great universal expositions. Here is the harsh opinion of a French economist writing in 1902: "More and more, universal expositions have lost their original character and become enterprises for pleasure. Interest in industry and commerce is only a pretext, and amusement is the aim. On the other hand, the specialized expositions remain serious and many of them are followed by important discoveries. . . . So it was with the electricity exposition which brought important progress for the telephone. . . . The universal exposition has had its day and no longer answers real needs; it will now be necessary, in order to benefit a country's commerce, to have recourse to other, less expensive but more productive, means."[31] In addition, he added, the creators of scientific knowledge were establishing more institutionalized forms of cooperation.

Other observers were more attentive to the cultural metamorphosis this crisis represented. In fact, the ever-larger part played by the logic of spectacle in the "universal exposition" formula brought out for the first time the tension between entertainment and high culture—but it also stimulated thought about the popularizing of science through pedagogy. It was in France, "mother of arts and letters," during the expositions of Paris—that "capital of the nineteenth century," in Walter Benjamin's expression[32]—that this tension began to be felt as early as 1889, year of the centennial of the Revolution and the inauguration of the Eiffel Tower. It would intensify in the 1900 expositions, which remained open for 205 days. Apart from the numerous games and attractions, there were cycling and automobile races, marksmanship contests, aeronautic contests, and fencing matches. Sporting competition became a popular attraction, fifty years after the exposition of London had launched the aristocratic "America's Cup."

In 1889, entertainment was internationalized, with the major attraction being Colonel Cody, alias Buffalo Bill, and his troupe of "redskins." Here is how a journalist of the magazine *L'Illustration* in May 1889 saw the effect provoked by the "Napoléon of the prairies":

> A painter would be disarmed by the ardent tonalities. As for all those seminaked beings, agile and galloping on their horses, whirling bareback on their mounts, they give a fantastic impression, an extraordinary fantasy. . . . How can the theater match such realities? If schoolchildren had to choose between a masterpiece of Corneille or Buffalo Bill, they would cry unanimously: "Buffalo Bill, Buffalo Bill!"[33]

Money as Medium

What is novel in the mode of production established by modern capitalism? This was a question that Marx, from the 1840s on, never stopped trying to answer: money is a medium, the agent of communication par excellence, the *perpetuum mobile*; its nature is to cross borders.[34] Time was to confirm these observations, which would outlive even political regimes that proclaimed themselves heirs to the author of *Capital*.

Indeed, Marx's analysis of commodity fetishism is not far from the one proposed by the anthropologist Georges Balandier on the eve of the third millennium. "Money," he writes, "expresses the essence of societies where almost everything can be translated into commodity terms; moreover, it informs a social and cultural universe where information is the energy indispensable to more and more numerous activities, and it designates perfectly the exchange relations in a world of communication and the rapid multiplication and intensification of exchanges of all kinds. It fully suits societies of this type: by the market, it regulates; by distribution, it hierarchizes; by investment, it grows. It appears as a generator of order. This occurs also on the terrain of the imaginary, in places where desire, fantasy, and play meet. The adventures of capital are converted into stories, fragments of myths and epics of a certain modernity."[35]

In an effort to combat this ascendancy of money, the International Association of Workers was founded at a meeting at St. Martin's Hall in London in September 1864. One passage of the organization's by-laws reads as follows: "The emancipation of labor, being neither a local nor a national problem but a social one, embraces all countries in which modern society exists, and requires, for its solution, the theoretical and practical cooperation of the most advanced countries. . . . The Association is established to create a central point of communication and cooperation between workingmen's associations in different countries aspiring to the same end, to wit: mutual assistance, progress and the complete liberation of the working class. The General Council will function as an international agent between different national and local groups, in such fashion that the workers of each country may be constantly informed of the movements of their class in other countries. . . . To facilitate communications, it will publish an international bulletin." (This was to be *The International Courier*, published by the London Branch of the IAW.) A new collective actor had appeared on the international scene.

But this first profession of generous faith in the international character of the demands made by the new class thrown up by industrialization would quickly run into the snags provoked by national and local differences, and other kinds as well. This pioneering organization, which tried to weld together the working classes of Europe and the United States, would disappear in 1874 after a triple failure: the Franco-Prussian War, the defeat of the Paris Commune, and divisions among the different components of the working-class movement (Marxist and non-Marxist). *The*

Communist Manifesto of Karl Marx and Friedrich Engels, first published in German in 1848, had already been through a dozen different editions in Germany, Great Britain, and the United States, as well as several editions in English and in French, and versions in Russian, Polish, and Danish. The Second International saw the light of day at the Congress of Paris in 1889 under the auspices of the social-democratic current.[36] The Third International was founded in Moscow in 1919. All these stages merely signaled the profound divisions in the working-class movement. And, in one way or another, they were all pregnant with meaning for the evolution of different modes of representation in the international space—and, within it, communicational space—opened up by the capitalist mode of production.

Very few of the participants at the first meeting of the International Workers Association in London would have been able to foretell the destiny of that phrase which made of the workers' societies in a few European countries the beacons of emancipation: "Emancipation of labor . . . requires the theoretical and practical cooperation of the most advanced countries." This disinterested offer by the workers of the "modern societies" also held surprises. The occasions would not be lacking for a rhyming of universalism and ethnocentrism.

Public Service: An Alternative to Jacobin Centralism

What questions were posed about the systems of communication under construction during the second half of the nineteenth century, in the milieus of this new form of social network that was nascent socialism? We may note at least three.

The first concerns the definition of "public service," which in turn refers back to the much vaster issue of the nature and role of the state. This was a question that divided the workers' movement between partisans of an "evolutionary" conception of social change and those who espoused a revolutionary vision and for whom the destruction of the state machine put in place by the industrial bourgeoisie was seen as a prerequisite for any change. This debate was formulated for the first time in a report on the public service drafted by the Belgian socialist César De Paepe (1841–90), who advocated an "antiauthoritarian communism," and it was presented before the International Association of Workers in 1874. The notion of "communication" in his contribution was limited to the railways, the post and telegraph, and the road system.

The author distanced himself from the two great currents of ideas that dominated economic discussion of the day: laissez-faire, which tended to abandon public services to private initiative, transforming them into businesses, and the interventionist school, which tended to place them under the control of the state. He noted also that the positions were less sharply distinct than was customarily supposed in opposing socialists and economists of the status quo: "In both camps, we find, alongside bourgeois economists who are partisans of the maintenance of the capitalist and working classes, socialist economists who want to abolish bourgeois and proletarian."[37] For De Paepe, public service was based on the acknowledgment of the utilitarian character of an activity, a usefulness that "wouldn't exist if one waited for private initiative, either because it would be diverted from its true destination or because it would constitute a monopoly that would be dangerous to abandon to private interests."[38] Thus "public service should be doubly public: 1) in that it is accomplished by the direct or indirect cooperation of all; 2) in that it has for a direct or indirect purpose the benefit of all. The true public service is thus at once public both by its subject and its object."[39]

De Paepe here revives a polemic begun by Proudhon in his book *Des réformes à opérer dans l'exploitation des chemins de fer* (Reforms to be adopted in the operation of railroads) and pursued by the anarchists. At first, and in this book particularly, Proudhon believed that the major and minor roads, canals, rivers, and so on should belong to the state and be maintained at its expense; as for the railways, the state should take charge of the construction of the lines and embankments and conserve eminent domain over the track; then, taking into account conditions relative to fares, the state should abandon the operation to individual companies destined to be transformed one day into workers' cooperatives. In another work, however, *Idée générale de la révolution au XIXe siècle* (General idea of the revolution in the nineteenth century), Proudhon removed one by one from the state all its prerogatives, and then suppressed it altogether as a useless mechanism. In contrast to Proudhon, De Paepe foresaw the dangers in abandoning public services—which "constitute for the most part a monopoly de facto, in other words, a privileged situation which favors speculation, [and] exploitation of the public—to companies, even if they are composed exclusively of workers," because, he asserted, "we must not forget that the modern capitalist aristocracy, too, came out of the Third Estate; let us not forget that before being what they are today, the great financial barons (or, if not them, at least their fathers or grandfathers) were workers, placed in a privileged position."[40]

In defining public service as "public at once by its subject and its object," in posing the question of citizen participation, in proposing a new conception of the double role of the local district and the state and in combining political decentralization and economic centralization, De Paepe was aware of offending both the Manchester liberals for whom the state should be reduced to the army, the courts, and the police, and, in De Paepe's own words, "the Jacobins of all shades [referring to the centralist tradition of the French Revolution] for whom the state is the be-all and end-all, the god Pan which gives life and motion to everything, for whom the state is the body social itself, and who do not understand that one may be born without a ticket of entry from the state, and depart this world without a state-issued passport."[41]

To our knowledge, this is one of the first critical reflections on the relation between public service, civil society (not reduced to its workplace activities), and the state.

Representation of the People

The second question that mobilized socialist circles is that of the creation of their own newspapers or periodicals. What model of the press should they adopt in order to reach a public wider than socialism's own circle of activists? How should they define themselves with respect to the mass press, with its criteria of commercial success and its different formulas for capturing a broad readership?

To answer these questions required them to ask about the dramatic force of the popular serial, and more generally, about mass culture as a culture of entertainment, which was just then being born. The subject was particularly thorny since Marx and Engels had, in *The Holy Family*, heaped anathema on the Saint-Simonian novelist Eugène Sue, author of *Mysteries of Paris, The Wandering Jew, The Seven Deadly Sins*, and so on. This serial production, which some condescendingly called "illiterate literature," was suspected of alienating the people. While the dailies of Catholic persuasion appropriated the serial formula, careful to offer the populace what it was accustomed to while "subverting" its contents, the socialist paper *L'Humanité*, founded in 1904 under the editorship of Jean Jaurès, relied only secondarily on this genre. It almost never offered unpublished *feuilletons*—which it could not afford in any case, given its perpetually precarious financial situation. Most of the very few original serial novels published before World War I in *L'Humanité* are, in the opinion of Anne-Marie Thiesse, historian of popular literature, "of mediocre composition or else their militant orientation is discreet." Which led her to the lapidary judgment: "The extreme poverty of the socialist *feuilleton* in the country which gave birth to the genre is a fact concealed by the diversity of original novels that appeared in *L'Humanité*."[42] The fiction section of the paper was fueled above all by a large number of foreign novels in translation: Russian novels (Tolstoy, Gorky), works by socialist Americans (Upton Sinclair, Jack London), detective stories

(Conan Doyle, Edgar Allan Poe), or popular stories, in addition to serialized realist or naturalist novels of the nineteenth century: Balzac, Flaubert, and above all Zola.

Thiesse concludes:

> This policy regarding the *feuilleton* had no great success with the general public. . . . Did it reflect the confusion of intellectual editors about a popular column they did not know what to do with? We may also see in it the absence of thought on the social problem of reading among the working and peasant classes. Between the attempt to make the masses appreciate classical culture and the temptation to turn to proven recipes of the commercial novel for the "People," French socialists could not find their way, any more than they knew how to create a product appropriate to the uneducated public that had to be won over to revolutionary ideas.[43]

Thiesse's verdict may be too summary, but it has the merit of signaling a tension that was to prove constant in the cultural strategies of forces claiming adherence to socialism, and not only in France. If we seek a more nuanced judgment, we must nonetheless recall the shrieks uttered by doctrinaire nationalists when they read the works of Tolstoy and other foreign authors. We may cite as an example an article that appeared in *Le Figaro* on July 29, 1892, under the signature of the novelist Maurice Barrès, champion of the national heritage and the fatherland. Entitled "The Quarrel Between Nationalists and Cosmopolitans," this profession of faith stigmatized believers in the nation as a contract or a voluntary association and denounced the invasion of cosmopolitan ideas, while defending poets belonging to the classical French tradition against Romantics who admired Tolstoy, Ibsen, and Maeterlinck.

Elsewhere, in czarist Russia, a third question gradually took shape: that of the relation of the press to the building of a revolutionary party. It was after 1895–96 and the famous strikes in St. Petersburg that a mass workers' movement allied with social democracy appeared and, with it, the need for a working-class press. This was the beginning of the Leninist doctrine of the press, for which a paper is not only an organ of propaganda and a collective agitator but also a collective organizer. This role fell to *Iskra* (The spark).

One of the most significant polemics on the role of the revolutionary press dates from 1901–2. Lenin's initiative, faithful to his conception of "democratic centralism," gave rise to numerous objections, notably on the part of L. Nadiejhdine and B. Kritchevski, who reproached Lenin for making of the movement's paper an instrument of unification from above. Refusing to convert themselves into "party inspectors," these two journalists, both active in the opposition to the czarist regime, confronted the future founder of the Communist Party with the idea of a press that would be the product of strong local organizations, the result of the "progressive march of the obscure daily fight, as opposed to propaganda." This would be, in short, a press that, unlike *Iskra* as they saw it, would take into account "the ordinary things of life."

This brought them a stinging response from Lenin: "All political life is an endless chain consisting of an infinite number of links. The art of the political activist lies precisely in finding and taking a firm grip of the link that is least likely to be struck from your hands, the one that is most important at the given moment, the one that most guarantees its possessor the possession of the whole chain." Anticipating the indignant reaction of these critics to his very directive proposals, he added a footnote: "Comrades, I call your attention to this outrageous manifestation of 'autocracy,' 'uncontrolled authority,' 'supreme regulation,' etc. [all terms that the two polemicists had applied to Lenin in their articles]. Just think of it: a desire to *possess* the whole chain!!! Send in a complaint at once. Here you have a ready-made topic for two editorials in your review."[44]

These prerevolutionary years were decisive for the formation of a certain conception of the use of the media: the means of mass communication as tools of agitation and propaganda.

World War I would bring the question of propaganda out of its revolutionary hideouts and raise it into an affair of state.

Chapter 3
The Invisible Management of the Great Society

World War I: Baptism of Fire

World War I was the first conflict that was referred to as "total." Not only did it take place at the world level but also, and more especially, it was a conflict in which political, economic and ideological warfare became as decisive as the operations on the battlefield. Inspiring the loyalty of citizens to the national cause became a task of top priority. Not only were more and more sectors of the national economy being called upon to contribute to the war effort, but the civilian population had begun to be more and more directly affected in their daily lives by this new form of confrontation. To the blockades and other embargoes were added mustard-gas attacks from the air. As the philosopher Walter Benjamin put it in a little-known text that he wrote to counter the theses of Ernst Jünger on the aesthetic of the new war and its warriors: "When the distinction no longer exists between civil population and the fighting forces—a distinction effectively abolished by gas warfare—the most important foundation of international law also disappears."[1]

Propaganda earned its first stripes as a technique of managing mass opinion, but also as a means of putting pressure on foreign governments. As a sample: two photos on the desk of an intelligence officer, one showing the corpses of soldiers being transported to the rear lines to be buried, and the second the remains of dead horses being sent to a factory to have oil and soap extracted. The officer composes the caption "Cadavers of soldiers leaving for the soap factory" and sends the photos off to the press.

No, this is not a scenario from Timisoara in December 1989 in the days preceding the fall of the Romanian dictator Ceauşescu. It happened in the spring of 1917 in London, in the offices of the Department of Information. The officer who tampered with the two photos seized from a German prisoner was a certain General Charteris. His objective was to persuade China to join the allied camp. The experts in propaganda and counterpropaganda at the time would recount after the war that the desecration of cadavers by the German army was intended to deeply offend the Chinese and their cult of the dead, and in fact this dispatch is said to have weighed heavily in their decision to abandon their neutrality.[2]

"A good propaganda policy probably saved a year of war, and this meant the saving of thousands of millions of pounds and probably a million lives," was the judgment of the London *Times* on October 31, 1918, eleven days before the armistice. But it should not be read necessarily as just one more example of official persuasion during the first global conflict, for this conclusion was in fact shared by many civilians and military people throughout the allied countries, and even by their adversaries, inundated as they were by tracts launched behind their lines inciting them to desert. One of the last bulletins of the eighteenth Army of the Kaiser's Germany reads:

> On the front of leaflet propaganda the enemy has defeated us. We have come to realize that in this life-and-death struggle, it was necessary to use the enemy's own methods. But we were unable to do this. . . . The enemy has defeated us, not hand-to-hand in the battlefield, bayonet against bayonet. No! Bad contents in poor print on poor paper had made our arm lame.[3]

The same admission was made by General Von Hindenburg in his memoirs, published shortly after the cessation of hostilities.[4]

For this weakness of the German propaganda front, commanded by a former primary-school teacher, Mattias Erzberger, from Budapesterstrasse 14 in Berlin, three explanations are necessary. First and above all, German propaganda appealed to reason, in an attempt to justify the attitudes of their compatriots. British propaganda, meanwhile, was addressed to the emotions, trying to incite indignation and revulsion. So when London put out news announcing atrocities committed by the enemy soldiery, photographs showing them engaging in pillage and the like, Berlin launched itself into long dissertations pointing out that only the United Kingdom's interest in liquidating its rival's industry had justified the war and explaining in great detail the historical and diplomatic reasons for Edward VII's policy of encircling Germany. When the Englishwoman Edith Cavell was condemned to death in 1915 in occupied Belgium by the German Army, and the public was revolted by this barbarous act committed against a woman (and a nurse to boot) accused of enemy intelligence, the only reply Berlin found to calm this emotional wave was to exhibit an article of international law. In contrast, the Germans did not succeed in scoring any publicity points on France's execution of one of their own spies, Mata Hari.

The second reason for this German weakness relates to the sheer quantity of propaganda deployed by the Entente forces. It intensified after August 1917, making its top priority to provoke the most desertions possible from enemy ranks. "This war is not your war": leaflets dropped by allied planes and balloons incited soldiers to revolt against Prussian militarism in order to institute a republic, and against officers gorged with food while ordinary troops faced privation. They promised that deserters would be well treated. The psychological offensive aimed at sapping the morale of the German soldiers was so effective that in 1918 the high command offered 3 marks to each combatant delivering to his superiors the first copy of a leaflet, 30 pfennigs for the following ones and 5 marks for a book. In May 1918, the officers gathered 84,000 leaflets in this manner, and in September more than a million. As to the actual number of deserters, the allied propaganda services after the war numbered them at forty to fifty thousand, immediately adding that in the pockets of many prisoners they found allied leaflets or pamphlets.[5] During the final allied thrust, the leaflets would give figures of German losses, notably U-boats, and Entente gains.

The final German handicap was that the dissension between the civilian government and the army chiefs of staff relegated to a low priority the creation of an office to coordinate the propaganda effort. Only military logic prevailed. Senior officers did not generally understand until too late the profound change implied by what was being tested, which would affect the very definition of war: its "total war" character.

The high command and the political authorities of the German Empire nevertheless realized the importance of film as an instrument of propaganda. It was on their initiative and thanks to the support of big banks that in 1917 the Universum-Film-Aktiengesellschaft, the famous UFA studio, was founded. Heart of the cinematographic effort on that side of the Rhine, it was to absorb the majority of existing companies, combining horizontal with vertical integration: production and distribution, from the manufacture of raw film stock to exhibition on the screen. The communiqué published on the occasion of UFA's founding announced: "We are happy to note that the opinion according to which film does not aim exclusively to entertain the public, but must respond to national needs, educational and economic, is becoming more widespread."[6] Nevertheless it was only after the war that the film industry took off, permitting Germany to become the second world producer, after the United

States, with the two countries' cinematographic industries interestingly displaying analogous structures, notably a strong concentration of, and intimate link between, finance and industrial capitals. The Hitler regime would transform this complex into a powerful propaganda machine.

The Army and the Media

The Allies succeeded, then, where their enemies had failed. In order to bring to fruition a strategy of persuasion, each of the great powers in the Entente set up its own structure; inter-ally coordination did not intervene until shortly before the armistice, and not always satisfactorily.

The United States created a Committee on Public Information, reporting directly to the president and composed of the secretaries of the navy and war and state and a journalist, George Creel.[7] This body, known as the Creel Committee, tried to mobilize the media in order to "sell the war to the American public" and thereby overcome the reluctance of the pacifists. The cinema was made to contribute. As the first mass medium, film had up to that point shown its potential as a tool of social integration in drawing to the theaters a strong contingent of poor urban immigrants, many of whom could neither read nor write, even in their own languages. The film industry began to produce propaganda films. But, propaganda or not, it turned to its own profit the weakening of European production. The war coincided with the first great wave of internationalization of American film, with films already amortized on the domestic market. It allowed the studios to vastly increase their production and to gain control of key positions in distribution and exhibition all around the globe. In Europe, the supremacy of French cinema collapsed. France lost not only the bulk of its export markets, but control of its own market. Production dwindled and theater exhibitors had to supply themselves more and more with foreign films, principally American ones.

In the history of the United States, the Creel Committee represents not only the first official propaganda agency, but the first bureau of governmental censorship. Established even before Congress could approve the declaration of war, it would prove particularly severe during the hostilities, on American territory as well as at the front, particularly in northern France. Two laws were passed to reinforce its powers, the Espionage Act (1917) and the Sedition Act (1918). The second, in particular, served to prevent the expression of any criticism of government policy. The severity of censorship toward transgressing journalists and newspapers drew such criticism that once hostilities were over, the government was obliged to soften it considerably. According to a report from the Gannett Foundation, not until October 1983, that is, until the Marines' intervention on the island of Grenada, did a military power "restore a kind of *de facto* censorship regime" as strict.[8] This regime was perfected and would become draconian during the Gulf War with the establishment of the "pool" system of coverage.

In Great Britain, the opposition parties showed themselves extremely vigilant in face of the government's inclination to censorship. Fearing that the government "spends money for propaganda in its own interest rather than in that of the country," they insisted on being represented on the committee in charge of domestic propaganda.[9] This pressure, combined with conflicts of jurisdiction among competing government departments, meant that, unlike the Creel Committee in the United States, the British structures changed several times. They did not acquire their definitive profile until February 1918.

Lord Beaverbrook, owner of the *Daily Express*, was made minister of information. In principle, this ministry had jurisdiction only over information destined for abroad. It consisted of several departments, including one, directed by Rudyard Kipling, in charge of American and allied opinion. Another was devoted to neutral countries, and a third, better known by the name of its headquarters, Crewe House, was given to Lord Northcliffe, owner of the *Daily Mail*, the *Evening News*, and the *Times*. He was in charge of propaganda for enemy countries. The German section of this department was entrusted to H. G. Wells, author of *War of the Worlds*. In the history of the British press, Northcliffe became the person who succeeded in saving the *Times* from bankruptcy, and also the one who "was capable of setting the world aflame in order better to light his posters" (in the words of the publicist Gardiner, one of his contemporaries). This reputation was largely acquired before the war took on mythic dimensions with the activities of Crewe House.

The presence of journalists and major newspaper publishers in the mechanisms of government propaganda merely underscored the influence acquired by the press since the start of hostilities in mobilizing the belligerent countries. As the prime minister, Lloyd George, had already expressed it in 1916: "The Press has performed the function which would have been performed by Parliament, and which the French Parliament has performed."[10] A governmental press office (the Press Bureau) did, of course, exist; it was given a negative function, that of preventing the publishing of news containing information useful to the enemy. Few papers, however, were prosecuted for violating security norms and none—except the *Nation*, which was briefly banned from circulation abroad in July 1917 because of its articles in favor of a negotiated peace—was suspended for expressing opinion contrary to that of the government, and this despite constant criticism from unions, socialists, and pacifists. Things went very differently in France.

Information Under a State of Siege

"At a time when a tragedy for which we are not responsible is about to be unleashed, it is wise to prevent certain discouraging impressions. Statistics from previous wars show that the better the weapons, the smaller the number of casualties."

"The inefficacy of enemy projectiles is the object of all commentaries: shrapnel bursting feebly and falling in harmless rain. The aim is badly adjusted. As for German bullets, they are not dangerous: they cross the body from one end to another without tearing the flesh."

The first quotation is not a misplaced one from January 1991 issued by the Iraqi Army when the troops of the coalition were getting ready to attack. It is an extract from the Parisian paper *Le Temps* of August 4, 1914. The second comes from *L'Intransigeant* on the seventeenth of the same month. Press censorship had been instituted through a series of measures that placed the press under extreme surveillance.[11]

On July 30, 1914, telegraphic dispatches were censored and the use of the telephone between one city and another forbidden. The second of August, a state of siege was declared over the whole territory and the military authorities could suspend or ban all periodic publications. On August 3, an "office of the press," reporting to the Ministry of War, was given the task of managing all military information. Two days later, a law on the "repression of indiscretions of the press in time of war" delineated in which cases information was to be controlled by the government or the military command: "Mobilizations, troop transport, the numbers of combatants, their disposition, the order of battle, the number of dead and wounded, changes in rank, along with all information or articles that comment on military or diplomatic operations that may assist the enemy or exercise an unfortunate influence on the morale of the army and population." On August 7, a circular was issued by the socialist Prime Minister René Viviani to newspaper editors: "The Government counts on the patriotic goodwill of the press of all political leanings not to publish any information concerning the war, whatever its nature or source, without its having been checked by the Press Office established yesterday at the Ministry of War."

In June 1916 an interministerial committee on the allocation of paper was created (it was to become the National Office of Paper in February 1918); its task was to set the quantity and price of paper according to the newspaper. In February 1917 the regulation of "formats and paginations" began.

The enforcement of these rules generated new administrative mechanisms of censorship. At first it was administered by the Ministry of War or its representative in each military region. The newspapers brought in their proofs and a military official of high rank read them all, indicating which passages were to be suppressed. Next, censorship was instituted at the Ministry of Public Instruction and confided to university professors, politicians, and officers. The suppressed passages were to be replaced by inoffensive texts so that blank spaces would not appear. Also in the early stages, instructions to the censors took the form of laconic communiqués. They later became more general in order to avoid misunderstandings. Here is one dated November 1917: "As for strikes, cut any information concerning the strike actions, all accounts or announcements of meetings in preparation for a strike, all polemical articles for or against a given strike, but allow articles of doctrine or news concerning the moderate economic demands formulated by the working class." In a work published in 1916, Gustave Le Bon made the first evaluation of what he characterized as the "method of silence" adopted by the authorities, and questioned the "uncertainties in battle stories." Regarding military motives, the psychopathologist wrote:

> The silence imposed on the press seems to have been caused partly by the fear, which is comprehensible, of the influence that public opinion might have exercised on the course of operations. In 1870, in the Franco-Prussian War, it was opinion that imposed on the campaign its most disastrous acts. . . . In the Spanish-American War, it was once again opinion that demanded that the fleet be sent to Cuba.[12]

In 1915, the government created an Information Section (SI), whose director was André Tardieu. He organized the publication of war chronicles in a *Revue de la vie du front* (Review of life at the front), and issued communiqués three times a day. The army set up an Office of Military Information (BIM) that accredited war correspondents.[13]

On the foreign front, the Great War was above all for France an occasion to take stock of the lag of its diplomatic corps in the realm of "means of intellectual action abroad" (in the expression of one writer in 1917):[14] Germany had taken a significant lead in the international book market; dissemination of French technical literature was ever more in decline; and book production and organization of the domestic market was also weak in France. In recent years, Germany had published an average of 34,000 works, against 12,000 for Great Britain and 10,000 for

France. Increased presence of experts abroad, the organization of international scientific conferences, as well as tours by great orchestras (such as the Gewandhaus) and the endeavors of numerous associations of expatriate Germans had all given the Kaiser's empire a collection of networks of cultural influence that France's Alliance Française, created in 1883, could not match. It was the realization of shortcomings of this kind that inspired the foundation, during World War I, of the *maison de la presse* by a government that was careful to incorporate press and publishing professionals into its policy of information disclosure. It was on this body's initiative that the first study was performed, by concerned professionals and the state, on the organization of the international distribution of French books. The *maison de la presse* soon had its agents in diplomatic offices. At the end of the war, in the spring of 1918, a special committee was born under the aegis of the Ministry of Instruction and Fine Arts to orient artistic propaganda abroad. In addition to official bodies such as the *maison de la presse*, this committee included organizations such as the union of high-fashion designers!

Clearly, World War I gave rise to interesting reflection on the balance of forces in international culture. The following comment, for example, dating from 1917, questions the place of French culture after the end of the war:

> We may admit that for a number of years Germany has acted, on account of its population stock and its exports, on the "content" of the peoples with whom it has relations, and that it is because it has wished to act on their "form" that it has raised almost the whole universe against it. It would be difficult for postwar France with a lesser population to be in a position to exercise a pronounced "material" action abroad: it would not have enough men both for itself and for its colonies. On the other hand, it is assured, if it wishes, of having a great effect on the "form" toward which the world after the war is tending.[15]

Demobilization

The day after the armistice, the information apparatuses set up during the war were dissolved. Mired in a political and economic crisis, the Weimar Republic abandoned propaganda efforts. This short postwar period saw an extraordinary flowering of arts and letters, but also new forms of publishing. It was the golden age of modern photographic journalism. In all big cities illustrated papers appeared, of which the most celebrated were *Berliner Illustrierte* and *Münchner Illustrierte Press*, which each ran to two million copies and were on everyone's doorstep. A host of celebrated photographers were on their staffs. In this period of exceptional effervescence, there was also a host of attempts at the social appropriation of new techniques of communication (cinema, photography, and radio). While in a number of places the workers' parties were extremely distrustful toward cinema—in Switzerland, the social democrats even proposed a boycott of this frivolous form of distraction that turned the workers away from the tasks of education—the German Willi Münzenberg published *Erobert den Film!* (Go conquer film!): not content with circulating films, he helped workers to produce their own, later repeating the experiment with photography and creating a network of worker-photographers.[16] In 1930 the playwright Bertolt Brecht, drawing in his turn on this philosophy in search of a horizontal communication, proposed changing the use of radio in a famous text that 40 years later would serve as manifesto for the *radios libres* movement against the state monopoly of the airwaves in France:

> Radio could be the most formidable instrument of communication that one may imagine for public life, an enormous system of channeling, or rather it could be, if only it knew not just how to broadcast but also how to receive; not only make the listener hear, but make him speak; not isolate him, but put him into relation with others. It will be necessary for radio, abandoning its role as supplier, to organize this supplying of listeners by themselves.[17]

The new democratic spirit manifested in the German press was brusquely killed off by the advent of Hitler in 1933. The editors of the illustrated papers began to be chosen on the basis of their loyalty to the regime. The effort was taken up by the French magazine *Vu*, founded in 1928, and by the American *Life*, founded in 1936.[18]

It was doubtless Great Britain that drew the most intelligent lessons from its experience of "information" during the first modern propaganda war. In 1926 the government created the Empire Marketing Board (EMB), whose mission was to publicize products of the empire using the whole range of media. This body rapidly became the first sponsor of the film documentary movement. In addition, it was John Grierson, one of the directors of the EMB, who, proposing a plan of action for "projecting England" (the title of his monograph), hastened the creation of the British Council and its network abroad.[19]

The British foreign information apparatus after the war was driven by economics. That of France continued to be driven by high culture and luxury goods. Here is a caustic judgment on this foreign policy between the wars, made in 1942 by Nicholas Spykman, one of the American pioneers of geopolitics, at a time when it was a question of countering the Axis's powers in Latin America and evaluating the forces needed to do it:

> France is a source of intellectual and artistic inspiration for the educated classes of both Spanish and Portuguese America, and it has needed very little effort to keep this favored position. Paris fashions and French luxury goods have met little competition in their appeal to the preferences of the Latin American buyers. With the "Alliance Française" operating in most of the capitals and a limited number of visiting professors lecturing before Latin American audiences, the French have kept the cultural situation in hand, but the results outside the luxury trades have been economically insignificant and politically inconsequential.[20]

The Locus of the New Power

The United States abolished the Creel Committee in 1919. The new winds of isolationism did not favor a strategy of official information directed abroad. This withdrawal did not mean a retirement from the "propaganda war," however, so insistent was the fear of the "Bolshevik menace." James Aronson, former journalist on the *New York Herald Tribune*, went so far as to assert that while "the date of the opening of the Cold War is most commonly set in 1946 . . . an excellent case can be made for fixing the date as March 3, 1918, the signing of the Brest-Litovsk Treaty, when the Soviets effected a peace with Germany and refused to continue in a war which the people of Russia had rejected. The 1918 date, in any case, marked the origin of the journalistic Cold War against Communism."[21] Proofs of this assertion are not lacking. The FBI under J. Edgar Hoover began to infiltrate unions, associations, and leftist groups, while the Department of Justice created a special publicity bureau with the mission of spreading stories of plots hatched by agents of Moscow and the "Reds" to overthrow the government in Washington. The attorney general, Mitchell Palmer, became a hero as a result of the raids bearing his name: in one single night, he had more than 4,000 alleged Communists arrested in 33 cities. Immigrants were deported and xenophobia inflamed passions. The trial of Sacco and Vanzetti and their execution in the electric chair in 1927 became the symbol of judicial error provoked by the pressure of public opinion stirred up to a white heat.

The major lesson the American government drew from World War I was strategic. In the course of the conflict, a technological leap forward had occurred: the development of powerful radio transmitters and listening stations, the coding of messages, the perfecting of mobile communications with cars and airplanes, radio navigation systems; and in 1915

Germany had taken the initiative of broadcasting news bulletins by radiotelegraph on the war operations and these were picked up by the foreign press. Naval operations had demonstrated the supremacy of the British radiocommunications industry and exposed the shortcomings of the industrial organization of that sector in the United States. With the British Marconi firm, the British Empire exercised a near-monopoly on wireless communications. Therefore the U.S. Navy, after 1919, asked the government to coordinate the efforts of large American companies capable of exploiting the new technologies of transmission classified as "strategic equipment." This concertation, which gave rise in 1920 to RCA, the Radio Corporation of America (originating from the takeover by General Electric of the American branch of Marconi), defined the areas of competence of each of the three great firms in the sector: AT&T would have exclusive rights to telephone and radiotelephone services, as well as the right to manufacture transmission equipment; trans-Atlantic services were reserved to RCA; General Electric kept wireless telegraphy and the manufacture of receiving equipment. The statutes of the new company growing out of Marconi featured three clauses: 1) its director must be American; 2) foreigners could not own more than 20 percent of the shares; and 3) a representative of the White House must sit on the board of directors.[22] In 1921, Westinghouse joined this trio.

Another pole of the development of American power in electronic communications was International Telephone and Telegraph. This corporation, whose mother company was founded in 1920 and operated the underwater cable between Cuba and the United States, acquired, five years later, the foreign subsidiaries of AT&T. Very early, ITT combined services (telephone, telegraph, and cable) and manufacturing. The acquisition in 1927–28 of All American Cables and Radio, founded in 1878, initiated the construction of its world telecommunications network. ITT World Communications began in 1926, soon followed in 1929 by Press Wireless, which would found agencies in the Philippines (1937), Uruguay (1942), and Brazil (1938). In the late 1920s, ITT supplanted British companies in South America by taking over, most notably, the United River Plate Telephone Company, the second largest company of the subcontinent, based in Argentina. In 1928 ITT Comunicaciones Mundiales SA of Buenos Aires and Santiago de Chile were inaugurated, as well as ITT Comunicaçãoes Mundiais SA of Rio de Janeiro. The American firm would later make Buenos Aires its regional headquarters. In its strategy of internationalization of its manufacturing and sales subsidiaries, ITT chose as priority territories Latin America and Europe. One index of its progression is the cascade of the "Standard Electric" subsidiaries it founded: Brazil (1937), Chile (1942), Mexico (1953), Venezuela (1957), Ecuador

(1962), Colombia, Jamaica, Panama (1963), Bolivia, Peru, El Salvador (1968).[23] On the European front, ITT would take control, after 1926–27, of the German subsidiaries of General Electric and of Phillips, and make a vigorous entry into the French telephone sector by taking control of Thomson-Houston. The takeover of the subsidiary of General Electric alone gave it a dominant position in the German, Austrian, Dutch, Danish, Swiss, and Turkish markets, as well as in the Balkans. This ubiquity of the firm would make it the paragon of multinational telecommunications enterprises, well before the concept was born.[24] The interlocking of its commercial and industrial interests would lead it to meddle directly in internal political affairs, and to show its sympathy explicitly for certain regimes. Its ambiguous relations with the Nazi regime during World War II is one of the darkest pages of its history. At the pinnacle of its expansion and power, it would not hesitate to plot in 1970 to overthrow the elected Socialist president of Chile, Salvador Allende.[25]

In 1926, RCA, whose original mission was to assure transoceanic telegraph services but which meanwhile had specialized in the manufacture of radio sets, bought up the station WEAF inaugurated by AT&T in 1922 in New York (until 1919, radio being considered a weapon of war, private broadcasts were banned everywhere). This was the first link of what was destined to become one of the three great radio (and later television) networks of the United States, NBC (National Broadcasting Corporation). Its original mission was "entertainment, information and education, with emphasis on the first feature—entertainment." Advertising was amply authorized. That same year, in 1926, in Great Britain, radio became a true public service with the creation of the British Broadcasting Corporation (BBC), after the purchase by the state of six private companies that had also begun to transmit in 1922. Advertising was forbidden on the airwaves. In France, where the first regular broadcasts took place in 1921, the radiophonic model that prevailed until the eve of World War II was hybrid, with certain "tolerated" private stations on which advertising was authorized, as well as public stations. (In 1940, when the German occupation began, the tolerated private broadcasters would be requisitioned.)

In March 1927, the Federal Radio Commission, created by the U.S. government to regulate air traffic, started work. (In 1934, it would be replaced by the Federal Communications Commission, the FCC, with regulatory power over telephone and telegraph as well as broadcasting.)

While Washington did not judge it necessary to elaborate an international cultural strategy, the film industry of the United States wasted no time pursuing its conquest of foreign markets. Hegemonic at the end of the war, as much in the Americas as in Europe, Hollywood—aided by its hatred of protectionism—assured itself of a strong position through the purchase of cinemas, the control of distribution, and the organization of local production. With the coming of talkies in 1926, the fight for supremacy on the world market shifted to the area of patents on sound systems. American patents (RCA and Western Electric) fought it out with German ones (Tobis, AEG, Siemens). In July 1930 the "Paris accord" put an end to this patent war; the world market was divided into zones for the sale of equipment. Each group obtained a zone of exclusive influence, but an open market was reserved where competition could operate freely. As a measure of the fact that film too was increasingly an industry, this agreement is a replica of the one reached in 1907 for the electrical industry when General Electric and AEG decided to divide up the globe and avoid harming each other in the penetration of foreign markets.

In 1930, American industry exported four times more than its European counterpart, and the process of concentration transformed its structure: the bulk of the production was in the hands of five majors (Paramount Pictures, Metro-Goldwyn-Mayer, Twentieth Century Fox, Warner, and RKO). But the introduction of sound film, where the question of language again became crucial, also permitted new countries to create or to consolidate their national production; this was the case in Italy, Czechoslovakia, Sweden, Poland, Switzerland, Mexico, and Argentina. American industry tried to resolve the difficulty through post-synchronization, subtitling, and the production of multiple versions in the languages of different countries.[26] The international presence of American film as a link between diverse national markets gave rise to the first polemics on the industrialization of culture—even more so since many European directors, scriptwriters, and actors were brought over by Hollywood producers to exercise their talents.

The Sources of Empirical Sociology

Peacetime brought back moral questioning of the purpose and means of propaganda, prompted by the publication of revelations by former propagandists seeking to repent, and the boasts of others who embellished the horror stories they had fabricated. Some isolated voices tried hard to combat the inflated reputation enjoyed by propaganda at a time when efforts were being made to determine the factors that had caused the fall of empire.[27] But whether they were for or against, the overwhelming majority did not contest the efficacy of the war of leaflets and communiqués. Both opponents and proselytes contributed to rekindling the idea of the miraculous power of modern techniques of persuasion. The debate was extrapolated to all media, and a conception of the inordinate power of mass media in fashioning people's minds made progress.

In this historical context was published the first representative work of what would later be known as mass communications research. It was entitled *Propaganda Techniques in the World War* and it appeared in 1927. Its author was Harold D. Lasswell, American inventor of the famous formula that was supposed to provide the sociological key to mass communication, the five Ws: "Who says What in Which channel to Whom with What effect?" This political scientist is considered by his peers as one of the four founding fathers of the discipline, the three others being Paul Lazarsfeld, the mathematician-turned-sociologist, of Viennese origin, and social psychologists Kurt Lewin and Carl I. Hovland.

The principal contribution of Lasswell's work was to bring out what the Prussian strategists had been unable to discern: Wherein lay the novelty of this first World War? His answer: in the necessity of "government management of opinion."

> During the war period it came to be recognized that the mobilization of men and means was not sufficient; there must be a *mobilization of opinion*. Power over opinion, as over life and property, passed into official hands, because the danger from license was greater than the danger of abuse. Indeed, there is no question but that government management of opinion is an inescapable corollary of large-scale modern war.[28]

Fascinated by what he termed the "effects" of propaganda, Lasswell hazarded an actual theory:

> Small primitive tribes can weld their heterogeneous members into a fighting whole by the beat of the tom-tom and the tempestuous rhythm of the dance. It is in orgies of physical exuberance that young men are brought to the boiling point of war, and that old and young men and women are caught in the suction of tribal purpose. In the Great Society it is no longer possible to fuse the waywardness of individuals in the furnace of the war dance; a newer and subtler instrument must weld thousands, even millions, of human beings into one amalgamated mass of hate and will and hope. A new flame must burn out the canker of dissent and temper the steel of bellicose enthusiasm. The name of this new hammer and anvil of social solidarity is propaganda.[29]

This exalted statement testifies to the belief in the omnipotence of propaganda that then reigned in academic circles in the United States. It was an era in which the mechanistic theory of stimulus-response in its primitive version prevailed. Propaganda's power of persuasion via the media left the public without defense, reducing it to the status of a passive receptacle of messages concocted by specialists of opinion. Lasswell's panegyric on persuasion also allows us to catch a glimpse of the theoretical frameworks of the time, whose points of reference have, in one way or another, been reinvested into thought on propaganda and its practice.

It is difficult to dissociate the first attempts at formulating a media theory from European thought since the late nineteenth century concerning the sociopsychology of public opinion. In the lead were the works of the Frenchman Gabriel Tarde. In the United States, this new discipline developed principally in the University of Chicago's sociology department. In this cradle of American sociology of communications under the sign of empiricism, the first effort to "measure attitudes" was born. In the dissemination of Tarde's ideas, one name stands out: Robert E. Park (1864–1944), specialist in the study of the role of the press in the formation of opinion, who for nearly 40 years exercised considerable influence within what came to be known as the Chicago School. In contrast to Lasswell, a man of the status quo, and in the manner of Charles H. Cooley, who is also considered as a representative of this school though he taught at Michigan, Robert Park belonged to a tradition of liberal criticism and never ceased to question the conflictual relation between media and democracy.[30] This sociologist who had reservations about the quantitative tendency of empiricism thought that there could certainly be no public opinion without substantial agreement among citizens. But he also thought there could be no public opinion free of disagreement. For him, public opinion presupposed public debate.

The direct influence of the philosopher Gustave Le Bon on the embryonic sociology of opinion is less apparent, though Lasswell was also the author of a work titled *The Psychopathology of Politics*. But the ideas of Le Bon, translated into fifteen languages, were too present in propaganda studies between the wars not to have left some traces in the Lasswellian conception of "mass,"as in the work of the American psychologist Leon Festinger, who was later to study "disindividuation" within the group.

Finally, the early theory of propaganda could not conceal the influence of the psychology of instincts represented by William McDougall, a physiologist of British origin who made his career in the United States. A representative of the biological paradigm, the author of *The Group Mind* insisted on the importance of instincts as determinants of human activities and spoke of the reflex reaction as a "direct affective induction." The affective state of the "soul" is merely the electric receptor as adjusted to perception.[31]

It is also striking to note that Tarde, Le Bon, McDougall—to whom we may add Sigmund Freud—are all major points of reference inspiring a

1927 book, *Psychologie de l'opinion et de la propagande politique* (Psychology of opinion and political propaganda) by the Frenchman Jules Rassak. But unlike Lasswell, this precursor of social psychology and active socialist adopted a critical position with respect to a number of these postulates in order to better question the growing contrast between the efficacy of the commercial press and the persuasive power of the ideals of socialism. The following observation is typical of Rassak's approach: "News announcing only facts has a propaganda effect much greater than political dissertations that smell like propaganda from miles away." In another passage, he relativizes the influence of the media: "It is much easier to spread tendentious news from abroad than from inside a country, because the reader is less able to verify it. This explains why the press is better able to provoke a war than to elect a municipal official."[32]

The Rape of the Masses

Not until the second half of the 1930s did some voices attempt a more systematic critique of the mechanisms of propaganda, in an environment haunted by the rise of Nazism.

With its ascent to power in 1933, Hitler's party had begun to concentrate in one ministry—the Ministry of Propaganda and Enlightenment of the People—all tasks concerned with "shaping minds." Its chief, Joseph Goebbels, proposed to organize this ministry into five sections: radio, press, cinema, theater, and general orientation of propaganda. The function of censorship was removed from the Ministry of the Interior. Information destined for foreign countries was removed from the Ministry of Foreign Affairs. The minister of postal services lost the monopoly of publicity on tourism, and commercial advertising was taken away from the minister of the economy. The artistic section of the minister of public instruction was confined to monuments, museums, music teaching, and popular libraries, while the direction of cultural policy resided henceforth with the super-Ministry of Propaganda. A Chamber of Culture was instituted; this was a corporatist organization subdivided into offices devoted respectively to literature, theater, music, radio, press, fine arts, and cinema. From that point on, membership in one of the specific offices was mandatory for all those whose activity was related to the production, reproduction, distribution, or preservation of cultural goods, from the seller of postcards and newspapers to the journalist, the painter, the director, or the writer.[33]

Two critical studies published at the time command attention. The first has become a classic. It is *The Rape of the Masses: The Psychology of Totalitarian Political Propaganda*, by the Russian immigrant Serge

Chakotin, a zoologist by training and professor of social psychology at the University of Paris. His work was published in French two months before the war and translated into English in 1940.[34] The occupation government had the original edition destroyed, and several months before its appearance, the French Minister of Foreign Affairs had enjoined its author to suppress "all passages disagreeable to Mr. Hitler and Mr. Mussolini" as well as the dedication "To the genius of France on the occasion of the 150th anniversary of the Revolution." With the law behind him, Chakotin refused to accede. The second work is certainly less well known. Its author is Robert A. Brady, professor of economics at the University of California, and it bears the title *The Spirit and Structure of German Fascism*. It was published in London in 1937.[35] (These two works had been preceded by the premonitory work by the Austrian Marxist psychoanalyst Wilhelm Reich, *Massenpsychologie des Faschismus* [The mass psychology of fascism], published by Sexpol Verlag in 1933 and quickly placed on the index of banned books by official Communism.)

The work of Chakotin has the appearance of a treatise on social psychology, presenting a vast fresco of the theories and doctrines on propaganda existing at the time. (The fresco would be completed in 1952 when the book was republished.) Only one work, published in 1946, rivaled the encyclopedic scope of Chakotin, a remarkable book by the Swiss author P. Reiwald, entitled *De l'esprit des masses* (Of the spirit of the masses).[36] Chakotin reviews the contributions of Tarde, Le Bon, McDougall, the behaviorists, and other precursors, passes Nazi propaganda through a sieve, and analyzes the mechanisms of Leninist propaganda. "It has been charged," wrote Chakotin in the French edition of the book,

> that these Russian practices are the same as those employed by Hitler. Yes and no. Yes, from the technical point of view of technique. Yes, in that in both cases the psychological base of affective propaganda is the same—drive number 1, or aggression. No, because with Hitler it was above all the element of fear that served to make the masses move in the direction desired by the state; in the USSR, the motor force is the opposite one from the combative drive—*enthusiasm*. In reality, what we call "elections" in the USSR are only a manifestation of what one is accustomed today to call "popular culture," employed to educate a people who will perhaps come someday to institute a true democracy. That is why the "elections" in the USSR are not a "comedy," nor classic "psychic rape," or demagogy, but a preparation, a prelude to a collective psychology.[37]

There is nothing astonishing in the fact that this author, so lucid on the "psychic rape of the masses by fascism and its heir—militant capitalism," and a friend of H. G. Wells, also practically worshipped professor I. P. Pavlov, famous for his experiments on conditioned reflexes, whom he acknowledged as his "great master" and whose theories of objective psychology he saw as the only ones able to provide an answer to the question: "What can be done to bar the road to evil propaganda?" And yet, in June 1922, the Soviet Revolution had restored Glavlit, the only institution of the old imperial order whose name was appropriated by the new regime. (It had been created in 1865 and attached to the Ministry of Interior Affairs.) By virtue of the law that instituted it, this governmental body was copresided by a representative of the organs of security and it was incumbent on him to ban correspondence, newspapers, magazines, films, books, drawings, radio programs, expositions, and so on that infringed the established norms (for example, agitation and propaganda against the Soviet authorities and the dictatorship of the proletariat, revealing state secrets, pornography, encouraging religious and ethnic fanaticism, etc.). At the level of the party, the department of Agitation and Propaganda (Agitprop) of the Central Committee was raised to the rank of supreme regulator of the flow of information. The 1922 decree was made more severe still in 1932, under Commissar Andrei Zdanov, three years before the great Stalinist purges (1935–38) and the beginning of the terror.[38]

The work of Robert Brady, on the other hand, is first of all a minute analysis of the corporate organization of Nazi society, and therein lies its principal merit. He analyzes the concept of propaganda as Nazi power had rendered it operational in the fields of arts, education, information, and science. He notes a double regime, a double legality, in the functioning of the system: on the one hand, firm state control in those domains of the formation of opinion; on the other, the predominance of the idea of self-organization in the world of business. He examines the differences and the convergences displayed by this conception of the Nazi "propaganda machine" by contrast with the conception implied by the strategy of public relations and business propaganda emerging in the United States. Accordingly, Brady mounts a veritable indictment against a prejudice he saw as common to the two approaches, according to which the audience is a "people who doesn't know what it wants" and who must be indoctrinated. He violently attacks works modeled on the psychopathology of Lasswell, which consider those who disagree with the system as "neurotic, pathologic, frustrated, or ignorant."

Brady's work undoubtedly contains a veiled message about what he judged dangerous for his own country: the rise of new forms of propaganda—because there too the political and economic stakes of the management of mass opinion were becoming apparent.

The Necessities of Fordism

"If we understand the mechanism and motives of the group mind, it is now possible to control and regiment the masses according to our will without their knowing it. . . . The conscious and intelligent manipulation of the organized habits and opinions of the masses is an important element in a democratic society. Those who manipulate this unseen mechanism of society constitute an invisible government which is the true ruling power of our country . . . it is the intelligent minorities which need to make use of propaganda continuously and systematically."[39] Thus the nephew of Freud, Edward Bernays, former member of the Creel Committee and one of the founders of modern public relations (to whom we owe the expression "engineering of consent") expressed himself in 1928.

Identifying the great society with the new postwar social order is not only a matter for university researchers. This imperative operated as well among managers and advertisers. As early as the 1920s, the founder of behaviorist psychology, John B. Watson, tried to elucidate the mechanisms and motives of mass psychology. (The first edition of his *Behavior: An Introduction to Comprehensive Psychology* was published in 1914.) Quitting the university to join the research team of J. Walter Thompson, the major advertising agency, in its search for the "mass consumer," he accomplished a considerable leap forward in the methods of commercial persuasion while modernizing the old instinctivist theory of stimulus-response. In contrast to McDougall, to these behaviorist psychologists the activity of learning appeared more important than the impulses of basic instincts. The individual and social human organism could be conditioned by appropriate treatment by playing on the dialectic of conditioned reflexes and intelligence. These advances in the technique of commercial propaganda took place in a period when the Fordist mode of production organization and labor control was being instituted in factories, and when managers were beginning to develop strategies for organizing mass consumption.[40]

Few countries asked as much as the United States of their apparatuses of mass communication, not so much because the media there had attained a more advanced degree of technological development than in most other industrialized countries, but rather because the media had, throughout this whole period, become the very cornerstone of a project of national integration. It was difficult not to subscribe to the analysis of sociologist Daniel Bell regarding the United States' slowness in constituting itself as a "national society." Neither the church, nor the party system, nor governing elites seemed to him to have succeeded in cementing national cohesion as much as had the media system. As he wrote in the French journal *Communications*:

> The element which has contributed to the amalgamation from the interior of our national society, since its apparition, outside of a few rare political "heroes" like Roosevelt, Eisenhower or Kennedy, has been popular culture. . . . The society, lacking clearly defined national institutions and a ruling class conscious of being so, congealed thanks to the mass media. Insofar as it is possible to establish a date for a social revolution, one could perhaps take the evening of March 7, 1955. That night, one out of every two Americans could see Mary Martin playing the role of Peter Pan on television. It was the first time in history that a single individual was seen and heard at the same time by such a broad public. This was what Adam Smith had called the "great society," but he could hardly have imagined to what degree this was true.[41]

At the end of the 1920s, however, the formation of the "great society" was seriously threatened by the Great Depression and its 13 million unemployed.

Communication as a Way Out of the Crisis

In 1933 the newly elected President, Franklin D. Roosevelt, launched the New Deal, which we understand as the rationalization of the state along with increased executive powers, modernization and regulation of the economy, the whole resting on the civic mobilization of the citizens.

For the first time in industrial society, the state, in search of a strategy to escape the crisis, summoned the "techniques of communication" to its rescue. The management of public opinion became an object of painstaking studies with operational aims. Roosevelt named "presidential agents" who traveled across the country to explain the administration's policy, give lectures, and take the pulse of national consensus. Nearly one and a half million trained propagandists wore their insignia of the Blue Eagle. In September 1933, more than 250,000 of these activists paraded through New York, escorted by 200 orchestras. Information on the behavior of voters and their attitudes regarding various policies or social problems became an ingredient of the art of governing.

The very notion of "attitude," introduced into the social sciences in 1918 by W. I. Thomas and F. Znaniecki, authors of a seminal work on the Polish peasant in Europe and the United States, and further developed by German experimental psychology, was becoming more refined.[42] To the pioneering definition of the two American researchers ("a state of mind of the individual toward a value")—who, incidentally, recognized their debt to Gabriel Tarde—the psychologist Gordon W. Allport added in 1935: "A mental and neural state of readiness, organized through experience, exerting a directive or dynamic influence upon the individual's response to all objects and situations with which it is related."[43] The structuring of American sociology was now under way.

Between 1924 and 1932, Elton Mayo conducted a series of studies of the largely female personnel in the workshops of Western Electric. This contribution to a budding social psychology, stimulated by the demand from industry, gave rise to a work called *The Human Problems of an Industrial Civilization* (1933).[44] In this work, Mayo rediscovers the importance of primary groups that form within industrial organizations. Observer participation at the workplace lost interest in measuring the effects of isolated variables and ceased limiting itself to "controlled experiments" and test situations, and instead began observing social situations understood as "systems of interdependent elements." Questioning the "latent functions" of workers' standard practices, spontaneous groups, and the organized behavior of the managers, this embryonic social psychology of the workplace aimed to better satisfy the needs of "human resources," to rally the workers to the company's objectives, and to better integrate them (using devices such as company newspapers, suggestion boxes, social services, professional training, pay schemes, and the like).

In 1937 Talcott Parsons published his major work, *The Structure of Social Action*, whose purpose, according to Parsons himself, was to create a unified social science on the basis of an empirical functionalism.[45] Empiricism was based on the closure of the object of study and on its division into an autonomous series of elements; functionalism proposed a vision of an "overall social system," insisting on the interdependence of all the elements of the system. At the center of this vision of society were the concepts of stability, equilibrium, and coherence. In the wake of Spencer, functionalism conceived the body social as constituted to assure its own "survival," in spite of all the centrifugal forces that might result in its disintegration. Each component of the social system plays a specific role, in order that a general equilibrium and the stability of the system as a whole may be preserved. In measuring each phenomenon from the point of view of its contribution to the maintenance of the social system's equilibrium, this research, which elevated functional laws into universal laws, took a given mode of organization of society as the natural frame of analysis and made it its ultimate horizon. Thus a contradiction could never be recognized as such, as a precursor to the emergence of another system. It would be defined instead as a "dysfunction" that endangered the equilibrium of the system.

Sociology thus became a form of social technology. It established a scale of values to define the more or less harmonious functioning of social institutions. Its object of study was, in the final analysis, to circumscribe the factors of disequilibrium in order better to control them. This explains why this functionalist sociology was to be so often the basis of therapeutic measures aiming to arrest the development of pockets of social dissent. This happened above all in the sixties, at a time when the alliance between the state and universities began to result in more and more commissioned research. It was also the moment when the early, prewar version of Parsonian functionalism acquired its specificity within the American sociology of mass communications, with researchers such as Charles Wright and Robert K. Merton. [46] This was indeed the early phase of functionalism, because in the sixties Parsons—still obsessed, to be sure, by the Hobbesian problem of social order—was to distance himself from the theoretical model of the living organism and draw closer to the cybernetic approach, which assimilated society to a self-regulating system.[47] By integrating into its theoretical framework disparate concepts such as motivations, normative orientations of actions, interaction, systems of expectations, modes of aggregation, and the institutionalization of actions and interactions, the Parsonian paradigm of sociology of action and systemic self-regulation demonstrated that one could no longer oversimplify the social, "considering it as a sort of linear extension of the individual, or inversely, as a simple matrix of the individual."[48] This complexity of the social escaped the sociology of mass communications of the sixties, blinded as it was by an inordinate optimism regarding its administrative mission. The evolution of Parsonian thought toward multidisciplinarity through contact with psychoanalysis, cultural anthropology, economics, and other fields contrasts with the disciplinary monoculture that characterized most studies in the empirical sociology of mass communications.

In the mid-1930s, opinion polls made their first appearance with George Gallup, former professor at the University of Iowa, who succeeded in predicting the reelection of Roosevelt in 1936. The first barometers of the opinion of the population were published just as the first research on the publics of the new radio networks began to stimulate interest in what really happened on the receivers' side. In 1939, the A. C. Nielsen company experimented with the first mechanical measure of the audience, the Audimeter, developed in collaboration with the Massachusetts Institute of Technology. It was first applied in 1942, with the appearance of the Nielsen Radio Index; the Television Index was inaugurated in

1950. By way of international comparison, it was in 1936 that Gallup set up operations in London, followed by Nielsen in 1939; and it was in 1938 that social psychologist Jean Stoetzel imported Gallup polls into France and founded the French Institute of Public Opinion (IFOP). In 1939, he founded the journal *Sondages*. The institute and the journal were reborn after the Liberation. But it was not until the great electoral battles of the Gaullist Republic in 1962 and especially in 1965 that the polling technique emerged from its semiclandestine role. And it was not until 1964 that the first large-scale inquiry on the television public was conducted, and the Office of French Radio-Télévision (ORTF) began to constitute a permanent panel of viewers for audience research.

The year 1937 saw the founding of the *Public Opinion Quarterly*, the organ of the American Association for Public Opinion Research (AAPOR) and the oldest journal of research into mass communications. It was edited by Princeton University's School of Public Affairs, and one of the members of its editorial board was Harold Lasswell. Many of the journal's contributors were closely involved with the new policy of opinion management that accompanied the New Deal. This journal, truly representative of its era, covered all the themes mentioned above in issues published between 1937 and 1941. As the United States broke with isolationism and World War II approached, international politics commanded more and more attention. Studies of German propaganda came back in force.

The first international organization in defense of professional interests in the field of advertising, the future International Advertising Association (IAA), was founded in 1938 in New York. Its main functions were to advance the general level of marketing throughout the world; to elevate the standards, practices, and ethical concepts of advertisers, advertising agencies, media, and allied services everywhere; and to encourage observance of the "International Code of Advertising Practice." This code had been issued in 1937 by the International Chamber of Commerce, established in 1920, and it became the major reference point for the formulation of codes of conduct in various countries. The idea that the profession had to defend the principle of self-regulation and oppose public regulation and controls was making its way. In Great Britain, this debate went back to the end of the nineteenth century. In the United States, the question of self-regulation, closely linked to the emerging notion of professionalism, had been debated in particular in the first decade of the century, first in local associations and then within the "Four As" (the American Association of Advertising Agencies, originally the Associated Advertising Clubs of America), founded in 1917. It was over the matter of these codes of professional ethics that the first international contact took place between representative bodies within the profession; in 1924, during the Great Empire Exhibition in London, the Four As met with the founders of the British Advertising Association (AA).[49]

But only the war allowed the United States to escape the Depression. In 1940, 15 percent of the working population was still unemployed, that is, more than 8 million workers. Between 1940 and 1945, the work force increased from 47 to 55 million, and more than 6 million would find work in the defense industries. The gross national product would more than double.[50]

Against the Martian Syndrome

In one way or another, manipulatory conceptions of the media left their imprint on the debates and the conceptual frameworks of the media between the two wars. The famous episode of the "Invasion of the Martians" may be understood as a kind of parable.

On the evening of October 30, 1938, millions of Americans were terrorized by a CBS radio program that described an invasion by Martians. The impresario was Orson Welles, who was dramatizing *The War of the Worlds*, the science fiction novel by H. G. Wells. The sociologist Hadley Cantril, to whom we owe an analysis of the program's impact, sums up the listeners' state of shock in this way: "Long before the broadcast had ended, people all over the United States were praying, crying and fleeing frantically to escape death from the Martians. Some ran to rescue loved ones. Others telephoned farewells or warnings, hurried to inform neighbors, sought information from newspapers and radio stations, or summoned ambulances and police cars. At least six million heard the broadcast. At least a million of them were frightened or disturbed."[51] The event created by Welles made it possible to test, for the first time on a large scale, the conditions of suggestibility and reciprocal contagion in a panic situation. From his interviews with those affected by the program, Cantril drew the conclusion that the best means of panic prevention was education. At the level of social perception, these scenes of unprecedented emotion, which were translated into thoughtless actions and gregarious crowd movements, were of no small influence in establishing the idea of the omnipotence of the new techniques of communication.

On a mental horizon dominated by preoccupation with the psychological effects of these social states, did any new conceptual vistas open onto analyses other than those springing from a conception of the receiver as a "suggestible" individual, a subject exposed to persuasion or alienation? Yes and no. No, if one adheres to a strict definition of media sociology.

Yes, if one looks elsewhere, that is, toward political philosophy and its examination of the relation between popular culture and its publics.

Herein lies the contribution of the Italian Marxist Antonio Gramsci (1891–1937), though it was not until well after his death that this contribution was appreciated—in fact, not until the end of the 1970s, when structuralist approaches to ideology, culture, and the media entered a deep crisis.

Marx and Engels had denounced the alienating function of popular serial novels. Gramsci, in response to numerous analyses of such novels (*feuilletons*) appearing in the early 1930s in French and Italian journals, tried to understand how the more and more "Taylorized" and disciplined activities of daily life had created the necessity for fantasies and dreams, the need for "illusion" and "daydreams." "It is necessary," he wrote, "to analyse which *particular illusion* (with respect to the novel, for example) is given people by serial literature, and how that illusion changes according to historical and political periods."[52] Because, he added, this literature for the people also contains "a basis in democratic aspirations." Regarding the success of foreign *feuilletons* in the Italian newspapers of his day, the question he posed was why "these readers of serial novels are interested in and are attached to their authors with much greater sincerity and more lively human interest than that shown in the salons of the so-called cultured people for the novels of D'Annunzio or the works of Pirandello?"[53] With the correlate: Why did Italy, as opposed to France, not produce this type of literature intended for the people, and why was it dependent on foreign production? All these frameworks of questioning only acquired their academic and political legitimacy when the receiver was rehabilitated as an active subject of the communication process, thus sealing the fate of theories of manipulation.

But the Italian philosopher had other areas of interest as well. The questions he formulated about the popular *feuilleton* take on their true meaning only in light of another question: the relation between intellectuals and the people and the role of the former in the production of consensus, or "general will." Further, he posed the problem of "mediations." To an oversimplified conception of bipolar divisions, he counterposed the complex range of systems of alliance and negotiations that figure in the establishment of the general will, or the process of the construction of hegemony, that is to say, the work of political, moral, and cultural leadership of a social group—a historic bloc—that penetrates to the heart of the body social, influencing its mode of living, its mentality, attitudes, and practical behavior. At the root of this reflection, a major concern: "One must fight economism," Gramsci wrote, "not only in historiography, but also and above all in theory and in political practice. One can and must fight in this domain by developing the concept of hegemony."[54] Here is an important conceptual and political opening that broke with the predominant conceptions of the time within an international workers movement that was inclined to see strategies of social change only in terms of economic struggles. With the concept of hegemony, Gramsci indicated that it was not sufficient to conquer the state and to change the economic structure in order to transform the old order; "culture" was a field where consensus in democratic societies was constructed on a daily basis, and that in this construction, the "intellectuals"—modern mediators—played an essential role. It was not until much later that Gramsci's analytical perspectives would inspire new critical approaches to the relations among culture, the media, and intellectuals. In the late 1970s, the weight of the media and industrialized culture in the production of consensus forced traditional intellectuals to redefine their relation to these apparatuses of mass culture, which were perceived as veritable "new organic intellectuals."

The struggle against economism also implied a new approach to the workplace. According to Gramsci, the emergence of a need to "dream with open eyes" is parallel to the institution of the scientific organization of labor (Taylorism) and the rationalization of production (Fordism). Accordingly, the factory is one of the places where hegemony is formed. It is also a pivot of his perception of the relation between the United States and Europe. Estimating the chances that the Fordist model would be introduced into European factories, Gramsci showed how the implantation of new methods of production was related to social changes that overflowed the four walls of the factory: changes in the type of state, in the relation between the sexes, in ethics—in short, in a "way of life."[55]

The United States was, in fact, in the process of becoming the mark of reference, and many an intellectual took a position on the question of "Americanism." Luigi Pirandello, a Nobel Prize winner in 1934 who embodied the intransigent figure of the creator in old Europe, went so far as to write: "Americanism is drowning us. I think that a new beacon of civilization has been lit there. The money circulating in the world is American and the realm of life and culture is forced to run after it."[56]

Examination of the influence of cultural models diffused by the United States was not, during these years, the exclusive fiefdom of old Europe. Thus, for example, the Peruvian José Carlos Mariátegui devoted a study to the genesis of "public instruction" in his country and criticized the inadequacy of the educational system with respect to national needs. Mariátegui not only analyzed the process of adoption of "American methods" following the Peruvian educational reform of 1920, but also retraced the history of the long preceding period, beginning in 1831, during which the ideas of French educators and thinkers had profoundly influenced Peru's leaders.[57]

The reflections on "the culture of Fordism" undertaken by Gramsci, writing from the prison where he had been locked up by Mussolini's Fascist regime, broke resolutely with the cultural vision of most contemporary European intellectuals. It was the hour of the Cassandras, when discussion went on at a lively pace about "the end of culture" and "the decline of the West," which was seen as succumbing to the blows of technical civilization. *The Decline of the West* was, of course, the title of the work of the German historian and philosopher Oswald Spengler (1880–1936); it was published in the early 1920s. And starting in 1926, in his Madrid journal *Revista de Occidente*, the Spanish philosopher Ortega y Gasset (1883–1955), author of *The Revolt of the Masses*, rose up against the culture exported by an America subject only to the laws of mass production and distribution and of technology. He saw this mass culture as a pseudo-culture destined to come inexorably into conflict with the high culture of the Enlightenment, of which the old continent had been the cradle and the guarantor. From the start, he denied that the United States had the capacity to "succeed Europe in ruling the world" or the power to fill "the gap of hegemony in a world which had lost its bearings of universality," because, he announced with great assurance, "New York is nothing new for us, any more than Moscow, by the way. . . . It is only our offspring."[58]

World War II would answer his claim by consecrating the advent of American hegemony over universality's representations, or at least representations of a certain universality.

Chapter 4
The Shock of Ideologies

Internationalization of the Airwaves

"The place of the artillery barrage as a preparation for an infantry attack will in the future be taken by revolutionary propaganda. Its task is to break down the enemy psychologically before the armies begin to function at all."

Thus wrote Hitler in *Mein Kampf.* In the course of World War II, this assertion would often prove relevant to the work of strategists seeking to overcome the reservations of generals regarding the psychological component of modern warfare. Nazi Germany prepared for the confrontation with a radically different conception from that of World War I. Specialists in geopolitics took note of this in 1938.[1]

That year, Washington entrusted six private companies—including the major networks NBC and CBS—with the production of radio programs for transmission abroad, with Latin America as a top priority. The American authorities finally rid themselves of the inertia into which they had been plunged by their isolationist policy and by a broadcasting system in the hands of a private sector whose major concern—as in many countries working under this arrangement—was anything but international coverage. There was much time to be made up. It was not until February 1942 that the government, on the edge of war, took over from the six companies and created a government radio station, the Voice of America.

The future Axis power had taken a considerable lead. As soon as it arrived in power in 1933, the National-Socialist Party inaugurated short-wave programs, in English and German, destined for the United States. Two years later, Italian Fascism began broadcasting in Arabic toward

Africa and the Middle East, obliging the British to move quickly to do the same. The Japanese Empire began to broadcast in English and Japanese toward Hawaii and the west coast of the United States; later, once the Sino-Japanese war had begun, it extended its sphere of action toward northern China and India.

In 1936 the Spanish Civil War showed the important strategic role that radio was called upon to play as a type of weapon. The public broadcasting facilities of the armies of General Franco initiated a large number of programs in Arabic from Tetuán, in an effort to dissuade inhabitants of Spanish Morocco from rallying to the Republican forces. The Republican radio in fact transmitted from Valencia in Arabic, as well as in French and even Russian, the latter destined for fighters in the International Brigade.

The Soviet Union, meanwhile, had proven itself a pioneer in the internationalization of the airwaves. In 1922, the Kremlin had what was considered the most powerful transmitter in the world. Faithful to the slogans of Lenin, this "newspaper without paper or borders" had in 1929 begun regular shortwave transmissions in German and French, and the following year in English and Dutch.[2] Before World War II broke out, the Soviet Union was broadcasting in more than ten languages and a multitude of dialects. But in transmitting power she was surpassed by Nazi Germany. The strength of the Soviet radio was based on the networks of its worldwide organization. At its Third Congress in 1921, the Communist International had published a programmatic document entitled *Theses on the Organization and Structure of the Communist Parties* that, in addition to ratifying the conception of the party as the vanguard of the proletariat, governed by the principle of "democratic centralism," proposed that its member parties profit from the Soviet Communist Party's experience in the press and in agitation and propaganda work, under both legal and clandestine conditions. The Comintern, as a centralized worldwide structure, soon revealed itself to be a fantastic instrument of "international communication," the national communist parties serving as relays and bases of support for the network. The first radio station established in Moscow in 1921 was in fact named Radio Comintern.

It was not until 1938 that the BBC—which was to be a catalyst in the fight against Germany—created a service in German and later began to broadcast in Spanish and Portuguese toward Latin America. Four years later, Broadcasting House in London was transmitting in sixteen languages, in addition to the regular broadcasts in English destined for the British Empire (the Imperial Service had begun in 1932). As for France, she had launched a colonial service in 1931, which was replaced in 1938 by Paris-Mondial. At this date, however, only one of the six transmitters called for in the initial project was actually in operation.[3]

On the eve of the world conflagration, Germany and England possessed more than 120 radios for each 1,000 inhabitants, while France had only 77; Italy, no more than 15. The United States had "taken off" with nearly 200 radio receivers per 1,000 inhabitants. On the other hand, the underequipment of Soviet citizens was flagrant—27 radios per 1,000 inhabitants—and this condition was not compensated for by group listening to "wire sets" connected to loudspeakers in sessions organized by the authorities, notably to better control their audiences.[4]

Where did the international organization of communication systems stand during these years? In 1920, an international conference on postal, rail, telegraph, telephone, and radio communication took place in Paris. The following year the Inter-Ally Technical Committee on International Radiocommunications prepared a conference to be held in Washington (1927) where it was decided to merge the International Telegraph Union (1865) and the International Radiotelegraph Union (1906). In 1932, during a conference organized in Madrid, this alliance gave official birth to the International Union of Telecommunications (IUT). (After World War II, this body, whose mission was to regulate all aspects of radiocommunications, would be integrated into the United Nations system in the same manner as the Universal Postal Union.)

From the beginning of radio broadcasting, states faced a triple problem: the dividing up of the spectrum of frequencies, the threat of "foreign aggression," and organizing the exchange of programs. In 1925, the international community set up an International Union of Radiodiffusion (IUR), with its seat in Geneva. (This would be one of the rare international institutions established in Switzerland to continue to function during the world war, under German hegemony and in the absence of representatives of radio stations belonging to countries hostile to the Axis.) The IUR and IUT together regulated the airwaves until the eve of the war. Numerous international congresses and committees, composed mostly of legal experts and diplomats, examined the best ways of assuring that radio would serve the cause of peace. The League of Nations had commissioned a report in 1931 on "all international questions raised by the use of broadcasting from the point of view of good relations between nations." This document served as a basis for discussion in the establishment of the Geneva Convention. This first international convention on foreign broadcasts, signed in 1936 by most of the member countries of the League, was tantamount to a pact of radio nonaggression. But reality proved to be stronger than agreements and conventions. In 1934, the only way imagined by the Dollfuss government to prevent Austrians from listening to Nazi propaganda broadcasts was to jam the airwaves. This was the first time this technique was used as a defense against this equally new type of "attack on national sovereignty."

A Geopolitical Vision of the Ideological Front

For the U.S. government, the main danger of the moment was the formation of a "fifth column" in the countries where German nationals lived. Ironically, the Third Reich tried to apply a doctrine formulated several years before by President Calvin Coolidge, a conservative and isolationist Republican, to the effect that national sovereignty extended to citizens and their possessions wherever they lived. By this criterion, the German nation included both subjects of the Reich (*Reichdeutsche*) and citizens of other countries if they were descendants of Germans (*Volksdeutsche*).

Relayed by the foreign branches of the shortwave station in Zeesen, not far from Berlin, Nazi propaganda aimed above all at rallying German residents abroad (estimated at 14 million people) and inciting them to form clubs and associations, or even mini-National-Socialist parties with a local *Gauleiter*—veritable pockets of subversion which, if need be, would prepare for invasions with acts of sabotage, create confusion, and spread Nazi ideas. The danger was considerable for the United States, which had to reckon with the presence of large concentrations of German descendants in the Americas—600,000 in Brazil, 150,000 in Argentina, and a strong colony in Chile—as well as with a long tradition of ties between the Prussian army and several national armies. It was in response to the threat of a Nazi "cultural front" in Latin America that the government in Washington was obliged to think for the first time about strategy in a wider framework than its own geopolitical interests. A flurry of decisions resulted.

The Roosevelt administration mobilized experts on public relations to study the best approaches to the Latin American countries. The State Department, in June 1938, formed a Division of Cultural Relations that worked—in spite of its general title—almost exclusively with countries of the southern continent. *Time* magazine and the *Reader's Digest* were enrolled in this task. The first issue (in Spanish) of a foreign-language edition of *Reader's Digest* dates from these years. Hollywood eliminated from its productions—which dominated Latin American screens—characters that risked offending the sensibilities of the inhabitants of the diverse republics.[5] Walt Disney was designated "ambassador of good will" and his animation studios in California appropriated popular figures from Mexico, Brazil, and the Andes. From this "good neighbors" policy came, notably, films such as *Saludos Amigos* and later *Los Tres Ca-*

balleros, and numerous episodes of comic strips.[6] Rotary and Lions Clubs placed their networks in the service of understanding among peoples. In this same framework, the government solicited the aid of private radio broadcasters, who offered their usual sponsors the chance to finance certain programs in Spanish or in Portuguese, such as President Roosevelt's fireside chats.

Nor did the United States neglect investments. World War I had inaugurated the era of financial hegemony for the United States, which, as holder of half the world's gold stock, had begun to become a massive exporter of capital. Between 1919 and 1930, its share in foreign investment rose from 6.3 percent to 35 percent.[7] A region already favored by investors, Latin America became under Roosevelt the chosen land of American firms. This preference would last nearly twenty years, since it was only in the 1950s that the center of gravity of direct foreign investment by the United States shifted to Europe.[8]

Old treaties were dusted off. A doctrine formulated by President Monroe in 1823 had asserted the necessity of preventing any extracontinental power from setting foot in Latin America, in the name of U.S. national security. But not until the eve of World War II did the United States approach its neighbors to set up a multilateral system of defense. Before that date, U.S. armed forces had left the field open to their future European enemies, with the German and Italian armies having privileged relations with their counterparts in several Latin American countries. The United States maintained relations only with the Brazilian and Peruvian navies, by means of a naval mission operated, by special Senate ruling, not by Washington, but by the host country! The first military mission was installed in Colombia in 1938. Only the zone of the Caribbean and the Panama Canal—"the American Mediterranean"—had, since the beginning of the twentieth century, been incorporated into the "American defense system."

The White House's cultural counteroffensive at the end of the 1930s was thus concomitant with a military strategy: it evolved in the framework of a geopolitical theory that established the concept of "hemispheric defense." The front that the geopolitical analysts of the time labeled "cultural," "ideological," or even "political" (the matter was never settled), earlier reduced to a minor role in international diplomacy, now acquired full recognition in world power relations. The front of culture, information, and ideology began to take its place in the art of war and in struggles for the conquest of a "hegemonic position." This latter term was dear to the American geopolitical analyst Nicholas Spykman, who defined it as the result of a combination of the military potential of a nation and the ensemble of factors such as "size of territory, nature of frontiers, size of population, absence or presence of raw materials, economic and technological development, financial strength, ethnic homogeneity, effective social integration, political stability, and national spirit."[9]

Psychological Warfare

During the course of World War II, the term "propaganda" was gradually replaced by that of "psychological warfare." As Harold Lasswell put it shortly after the surrender of Axis forces:

> The word originated and gained significance in Germany as the German s who were defeated in World War One began to look into the causes of that "collapse." . . . The vogue of the expression "Psychological Warfare" came in part from the rapid expansion of specialized psychologists in Germany, the United States, and in other Western countries. The psychologists wanted "a place in the sun"; that is, they were eager to demonstrate that their skills could be used for the national defense in time of war. Early in the Second World War a group of Americans translated some of the important German literature into English for the purpose of opening the eyes of the military to the usefulness of psychology, not only in testing for specific aptitudes, or in propaganda, but in considering every phase of the conduct of war under modern conditions.[10]

Lasswell himself had taken part in this popularization of doctrines on propaganda with his book *World Revolutionary Propaganda* in 1939.

Considering the trajectory of the term is useful above all for explaining the use made of it by the American military. The British—the BBC showed itself particularly effective in this type of operation—preferred the term "political warfare," the name given by an important agency, responsible for information disseminated abroad: the Political Warfare Executive.[11] The difference in terms is secondary, or at least so believed Ladislas Farago, who played a major role in bringing the notion of "psychological warfare" to the United States during World War II, and who saw the two terms as synonymously designating "that form of intelligence operations that uses ideas to influence policies. It deals with opinions and with their communication to others. It is organized persuasion by non-violent means, in contrast to military warfare in which the will of the victor is imposed upon the vanquished by violence or the threat of it."[12]

This common-denominator definition hardly masks the divergences that came to light not only in the area of doctrine but also in the field of jurisdiction—divergences that could by no means be hidden by the victory over Nazism. In this realm of definitions, while there are many terms, there are few available meanings. The endless debates on the difference between psychological warfare and information, propaganda and information, persuasion and communication, usually end in a scoreless tie, so fragile is the partition separating one from another, particularly in time of war. On this changing terrain, a definition is useless if it does not refer to uses.

From this point of view, World War II was without a doubt the first full-scale laboratory of the modern sociology of mass communications. Not only all the famous names of the era but also those of the future, all linked in one way or another to the fate of the discipline, took part by serving the cause of psychological warfare: Leonard W. Dobb, Carl I. Hovland, Alex Inkeles, Morris Janowitz, Joseph I. Klapper, Harold D. Lasswell, Daniel Lerner, Leo Lowenthal, Lucian W. Pye, Wilbur Schramm, and others, a list to which one should add many university professors of all persuasions, such as the anthropologist Clyde K. Kluckhohn or the philosopher Herbert Marcuse. All of them worked within the new structures created at that time.

The United States entered the war on December 7, 1941. At the time, they possessed only two agencies for conducting propaganda outside their territory: the CIAA (Office of the Coordinator of Inter-American Affairs) and the COI (Coordinator of Information). Founded in August 1940 and placed under the presidency of Nelson Rockefeller, the CIAA worked in close collaboration with the State Department, but its official mission was confined to Latin America. The COI, established in July 1941, was entrusted with covering the rest of the world. In June 1942, the COI gave way to the Office of War Information (OWI) and the Office of Strategic Services (OSS), with no clear demarcation between these agencies. It was not until March 1943 that the White House clearly defined the missions of the OWI and the OSS; the OWI was to orient information directed abroad and engage in "overt propaganda," while the OSS was placed in charge of "covert propaganda." Sociologists and psychologists worked within the OWI or the OSS. Many became advisers to the Voice of America.

The National Security State

In the immediate postwar period, a multitude of authors published works in which they attempted to draw lessons from their own practical experience. In a fascinating book, Clyde Kluckhohn recounted how anthropology had been pressed into service in constructing a coherent strategy of psychological warfare against imperial Japan—a strategy to demoralize troops while preserving a certain continuity in the social organization, with a view to allowing for a cultural transition.[13] Edward A. Shils and

Morris Janowitz published a prototypical study of the impact of Allied propaganda on the German Army's effectiveness in combat. The original feature of their work was its focus on the social structure in which the messages were received. Contrary to the widespread idea of propaganda as a panacea, they showed that it was only when primary groups (notably groups of friends) began to disintegrate that propaganda facilitated that disintegration. It was the fundamental indifference of German troops toward the millions of Allied leaflets and radio broadcasts that led the authors to examine basic military organization and its relation to the system of primary groups.[14]

The sociologist John Riley and the psychologist Leonard Cottrell drew up a survey of research and a list of recommendations that ran counter to the theories in force before the war. "The international models must replace the conventional stimulus-response concepts," they wrote, "if we wish to understand communication phenomena and to use this understanding in psychological warfare."[15] These conclusions were not shared by Carl Hovland and his team of experimental psychologists at Yale, who during the war had tested the effects of films illustrating the causes and aims of the war on American soldiers in the Pacific and on the European front, using as their basis the very schema criticized by the authors just cited. The Yale psychologist had even drawn a model: the psychodynamic model of persuasion. The persuasive message, according to him, was one whose properties were capable of altering the psychological functioning of the individual and bringing him to perform acts desired by the sender. A veritable "bible" of persuasion techniques resulted.[16]

Opinions were quite mixed regarding the impact on the development of social science of its enrollment in the war effort. One of the most severe critics was Paul Lazarsfeld:

> During the Second World War, most government agencies made extensive use of domestic communications research. This led to a multiplication of established activities rather than to a search for new problems and new methods, a kind of freezing at the pre-war level. At the same time, international communications research made its beginning. The concern with shortwave propaganda, especially from the German side, stimulated most of the early research and writing.[17]

But most of the evaluations did not gratuitously return to the past, for another war was on the horizon, one that would pit East against West for 40 years. Many "reenlisted" with their acquired baggage, this time in the service of the "Free World." Thus, in 1951, John Riley and Wilbur Schramm, both formerly with the OWI, published *The Reds Take a City:*

The Communist Occupation of Seoul, with Eyewitness Accounts. The tone was set.

In 1947, three years before the Korean War broke out, the OSS was transformed into the CIA (Central Intelligence Agency) and in 1948, the OWI gave way to the Office of International Information. The zeal with which the OWI was dismantled seemed to Wilson Dizard, a top adviser on governmental information in the early 1970s, "more than indecent." Some government officials viewed this body with suspicion, for in their opinion, the antifascist struggle had stuffed it with Roosevelt partisans and communist infiltrators. In the 1950s, at the time of the anticommunist "witchhunt," the U.S. Information Agency, the government agency for foreign propaganda, was subjected to four inquiries by the Senate Committee presided over by Joseph McCarthy. A purge ensued, and a number of suspect books were sent to the pulper.[18]

The law that instituted the CIA was known as the National Security Act. It legitimized the wartime institutions and made their priorities those of peacetime. Above all, it furnished the legal framework in which the exceptional mobilization of the war could be maintained, thus preventing a state of demobilization that would recall unpleasant memories of the crisis of the 1930s. The immediate postwar period had in fact plunged the country into a severe crisis of reconversion. The Marshall Plan—as well as the Korean War—stimulated a recovery. Unemployment decreased from 6 percent to 3 percent and the growth rate nearly reached 11 percent in 1950. In the view of analysts of the relation between military conflict and the health of the American economy, the return on the Korean War "investment" was remarkably profitable. As one of them remarked ironically: "Americans began to dream of a system that maintained a high level of military expenditure, independently, so to speak, of war."[19] One thing is clear: the recovery was worldwide and was followed by what one French historian called the "thirty glorious years," that is, the three decades in which the industrial economies enjoyed exponential growth.

The National Security Act sealed the permanent alliance between industry and the war-oriented state—an alliance without which the formidable takeoff of the aerospace and electronics industries would never have occurred. In presenting the Act before the Senate, the Secretary of the Navy, James Forrestal, summed it up as follows: "This bill . . . provides for the coordination of the three armed services, but what is to me even more important, it provides for the integration of foreign policy with national policy, the integration of our civil economy with military obligations; it provides for continual advances in the field of research and applied services."[20] And so the basis was laid for the synergy between private business and the Pentagon, industrial production and military re-search, university research and the needs of national security. This syner-gy had proved itself in World War II, and that is how ENIAC (Electronic Numerical Integrator, Analyzer, and Calculator) was born. ENIAC was a first-generation, large-scale electronic computer built—in top secrecy—by professors Eckert and Lauchly and their coworkers at the University of Pennsylvania for the Ballistic Research Laboratories of the U.S. Army Ordnance Corps.

As a measure of the qualitative leap in attitudes to research and devel-opment in government and military circles, it may be noted that in 1930, only 14 percent of the public and private budget for research and devel-opment was spent by the government; by 1947, the governmental por-tion of the national bill was 56 percent. As an expert at the Industrial College of the U.S. Armed Forces observed in February 1950: "The ex-pansion of the government's activities . . . is an index of the increased ability to administer these matters as systematic activities. The relation-ship between R&D and national security has long been known, but the scale of effort and the speed of progress have so increased that this rela-tionship now has an entirely new significance."[21]

Power Conflicts Between Disciplines

The Cold War precipitated a questioning of the concept of psychological warfare. "It has been my impression," L. S. Cottrell told the American Sociological Society,

> that far from being a clarifying concept that structures a field and guides action, to say nothing of research effort, the term "psycho-logical warfare" is ambiguous and leads to confused thinking and action. It lends itself to use for covering too much or too little, and to mistaken decisions as to appropriate divisions and responsibility in action programs as well as research. . . . Let me note here that I recognize the importance of the role of psychological theory and research methodology for the area we are discussing. [Neverthe-less] the strategic error the psychologists have tended to make, with notable exceptions of course, is that they did not rapidly move to supplement their own imaginations and skills with top quality competence in the other relevant disciplines, especially soci-ology, anthropology, social psychology and political science.[22]

In academic circles, efforts were made to find a replacement for the concept, or failing that, to sketch the contours of a new content. One scholar drew up an inventory of the expressions used in practice to desig-nate this polymorphous reality: "war of ideas, struggle for the minds and will of men, thought war, ideological warfare, war of nerves, political warfare, international information, overseas information, campaigns of truth, international propaganda, psychological warfare, war of words, indirect aggression, agitation, international communication."[23] One of the major concerns was to find a way of marking the differences between yesterday and today, between a meaning that had acquired its legitimacy under military rules, and another in quest of a more civilian orientation. As Murray Dyer put it: "In a given democratic society the answers to such questions must be given, [if] at all, by political, as distinct from military, authority even though military considerations may be a governing factor."[24]

The result of this debate over basic definitions made itself felt even before the 1950s were over. The term "psychological warfare" kept its place in the academic community, particularly among researchers coming out of experimental psychology, such as Carl Hovland and his team at Yale. But at the same time, concepts such as "political communication" and "international communication" made headway.

This dispute over concepts is explained in part—as Cottrell clearly implies—by the rivalry between disciplines and currents of thought, the stakes being the carving up of research territory and the tapping of considerable financial resources available to university research centers.

The competition between scientific paradigms is particularly visible in the evolution of theories on the effects of the media in the 1940s and 1950s. To a theoretical tradition based on the stimulus-response scheme and represented by Lasswell and his mechanistic conception of the process of communication, or by Hovland's adaptation of the behaviorist theory of learning, replied sociologists such as Paul Lazarsfeld. His work *People's Choice*, written in collaboration with Bernard Berelson and Hazel Gaudet, came out in 1944. These researchers tried to measure the influence of the media on 600 voters in Erie County, Ohio, during the presidential campaign of 1940. Their goal was to observe and evaluate intermediary elements that slipped between the initial and final points of the communication process and that had a direct influence on the "effects" obtained by a message.[25] This book gave rise to many others, not least the famous *Personal Influence*, coauthored by Lazarsfeld and Elihu Katz, published in 1955 but based on studies performed ten years before.[26] Taking up the conclusions of the first study, the two authors approached not only voting behavior but also the behavior of consumers in the market for consumer goods, fashion, and leisure activities, and looked particularly at their choice of films. In studying individual decision processes of a female population of 800 people living in a city of 60,000 (Decatur, Illinois), they rediscovered—as in the preceding study—the importance of "primary groups." They apprehended the flow of communication as a two-stage process where the role of "opinion leaders" was essential. This became the "two-step flow."[27] At a first level, there are those who are relatively well-informed because they are exposed directly to the media; at the second, there are those who have less contact with the media and depend on others to obtain information. From this first category are recruited the opinion leaders who retransmit information to the second group via interpersonal channels. What underlay these conclusions was a questioning of the theories and doctrines of "mass society" and its uniform effect on its inhabitants, as well as theses, optimistic or pessimistic, on the omnipotence of the media. The very notion of a massifying effect gave way to a hypothesis well summarized by Bernard Berelson in 1949: certain types of communication that refer to certain types of problems, addressing certain types of people who find themselves in certain conditions, produce a certain type of effect.[28]

The fact that in the evaluation of "media effects" one current of thought subscribed to the behaviorist hypothesis and another refuted it by invoking communication in two stages did not substantially change the matter. What separated the two positions was secondary, for they both began with the same presupposition: a self-contained individual, disconnected from any bond with society. For these heirs to the political philosophy of Hobbes and Locke, neither people nor the media themselves were socially situated—that is, situated in networks of divergent and contradictory interests, evolving in a structure in which they constructed themselves as subjects while at the same time being molded by it. Here lies the point of convergence where those who attributed a mythic power to the media joined those who, later on, relativized that power to the point of diluting it into the neoliberal principle of the absolute sovereignty of the consumer and his or her self-determination in making choices.

"International Communication": Discourse of Combat

Very few research reports on "international communications" in that era fail to sing the praises of the intellectual in the service of the "Free World." Here are some examples.

The first comes from a study by a famous Sovietologist, Alex Inkeles:

> Shortly after the end of World War II, the United States and the Soviet Union became locked in a large scale ideological struggle in which the weapon has been propaganda, the field of battle, the channels of international communications, and the prize, the loyalties and allegiances of men and women throughout the world. Undoubtedly, the most important aspect of this combat is its effect on

the minds of men, and the implications of such effects for national stability and international peace. The specialist on mass communication and public opinion has a major responsibility for studying those effects.[29]

The second example is an extract from a study made by Joseph Klapper and Leo Lowenthal in the course of their work as experts for Voice of America, the medium par excellence of the official policy of the United States:

The psychological warriors of the United States are today engaged in mass communication by press, film and radio. . . . This paper proposes to review the contributions of opinion research to one type of psychological warfare: specifically, its contribution to the evaluation of international broadcasting. Some of these contributions are adequate to the tasks at hand; others fall somewhat short of present needs. Such shortcomings, and suggested modes of overcoming them, will also be specified.[30]

When we remember the bellicose title of the work on the taking of Seoul by Wilbur Schramm—founder of the celebrated Institute for Communication Research at Stanford University (1956)—we are scarcely surprised to find in another of his works the following assertion of the necessity to "professionalize psychological warfare": "The world has progressed to a state in which self-preservation alone demands the most intense *psywar* pressure that a body of trained professionals commanding immense resources can bring to war."[31]

On the other hand, it is less easy to understand the promptness of Paul Lazarsfeld, chief of the famous Bureau of Applied Social Research at Columbia, in answering the call by offering his colleagues, in an issue of the *Public Opinion Quarterly* (Winter 1952–53), the outline of a new field of research on "international communications," whose construction seemed to him inseparable from the new political situation. In the same issue, Leo Lowenthal announced the birth of a "new discipline of international communication." Several months before, a subcommittee had even been organized within the American Association for Public Opinion Research to promote it, with Lowenthal presiding. "The relationship between practical policy and social science," wrote Lazarsfeld,

should be a two-way relationship. It is not only that we should contribute to the social sciences. This is imperative not merely for academic reasons but because, to a considerable extent, the national and international welfare of the country, as Lasswell points out, is tied up with the techniques of social research. The policy-makers should be joined by social scientists, not only because we can help them, but because the exclusion of the social sciences from the social events of the day impoverishes the social scientists who are themselves an important resource in a country. It is very much to be hoped that, in this sense, international communications research, because it is working in an exposed area, will contribute to the improvement of the relation between social sciences and those groups and institutions who are the actors on the social scene.[32]

This framework seemed to the dissident sociologist C. Wright Mills so narrow that he argued: "Sociology has lost its reforming push; its tendencies toward fragmentary problems and scattered causation have been conservatively turned to the use of corporation, army and state."[33]

It was in 1953 that President Eisenhower launched a resounding call to all active forces of the nation to defend freedom: "The struggle in which freedom today is engaged is quite literally a total and universal struggle. . . . It is a political struggle. . . . It is a scientific struggle. . . . It is a spiritual struggle. . . . For this whole struggle, in the deepest sense, is waged neither for land nor for food nor for power—but for the soul of man himself."[34]

On the technological front, the flow of investment from the Pentagon into research and development of new technologies of information allowed the computer industry to take off in those years. A report commissioned by the Organization for Economic Cooperation and Development (OECD), on the threshold of an era when international competition in this sector began to sharpen, read as follows:

> In 1959, research and development contracts worth almost a billion dollars were allocated to computer manufacturers (in the United States). This figure is comparable to total sales of computers on the civilian market in the same period, and it exceeds considerably all support given to the computer industry in other countries. Coinciding with these years during which this new and important industry was established, this policy no doubt had a much greater effect than any other national policy pursued at the time or since.[35]

A number of American historians of computer science agree on the fact that the Korean War was decisive in the expansion of "computer needs." They also agree that one contract in particular allowed IBM to outstrip all its rivals: the one it secured as lowest bidder for the Ballistic Missile Early Warning System (BMEWS), which gave this corporation the opportunity to conceive the first transistorized computer in November 1959.

During the 1950s, the continental defense network SAGE (Semi-Automatic Ground Environment) was built at the request of the U.S. Air Force. Each computer was linked to a radar unit that received data on aircraft flight trajectories. The transmission of information between computers, linked by telephone lines, made it possible to assimilate information from different origins and build up a continual data base in space. This network, which inaugurated "telecomputing," prepared the way for most of the innovations later used by all computers. According to historian Philippe Breton, SAGE no doubt "inaugurated the entry of man into the artificial worlds which he placed between himself and a nature more and more inaccessible in 'real human time' and served to stimulate industry, particularly IBM, toward the mass production of sophisticated and reliable computers."[36]

It was also under the aegis of the Department of Defense that the architecture of the first network of data transmission, the Advanced Research Project Agency Network (ARPANET), was organized in 1968. It was supposed to link together the calculation centers of American universities and, thanks to a satellite connection, link these centers with Europe via London and with the Pacific via Hawaii. Its official mission was to service all the projects by the federal government, and it would later serve as a point of reference for most of its counterparts in Western countries. Starting with its conception in the line of national security (the first experiments date from 1958), this system would retain the initial idea of a network of calculators linked in such a way that the transmission of numerical data could proceed by several different routes and that the whole would not suffer unduly from the destruction of one or even several centers of calculation.

The Space Race

Before the 1950s were over, another Cold War front opened: the space race. The Soviets overtook the Americans by launching the first artificial satellite, Sputnik, in 1957, and then in 1961 by putting the first man into space, Yuri Gagarin. President John F. Kennedy called upon his nation to redouble its efforts to reach the moon before the end of the decade. As one commentator explained, "The U.S., its psyche as well as its sense of security shaken by the prestige and military import of Russia's Sputnik, opened wide the fiscal thrusters and spent billions upon billions to catch up to and pass the USSR in space."[37]

That was the beginning of a frenetic dash that can only be compared to the arms race. The rush to the stars became a preeminent arena for winning the Cold War. The bridgehead was NASA (National Aeronautics and Space Administration), founded in 1957, with a mission to research and promote projects of space exploration. Its administrator proclaimed far and wide the new patriotic vocation:

> For the first time in the history of mankind the opportunity to leave the earth and explore the solar system is at hand. Only two nations, the United States and the Soviet Union, today have the resources with which to exploit this opportunity. Were we, as the symbol of democratic government, to surrender this opportunity to the leading advocate of the Communist ideology, we could no longer stand large in our own image, or in the image that other nations have of us and of the free society we represent.[38]

In 1962, five years after the success of the first artificial satellite Sputnik and the riposte by Explorer, the United States launched its Relay, Syncom, and Telstar satellites. The first of the Telstar series was the first "active" communication satellite, so called because it was equipped with amplifiers that magnified the signals transmitted toward the earth. It linked the United States to Europe for the first time. Also in that year, the U.S. government set up an institution whose mission was to exploit space technology: Comsat (Communication Satellite Corporation). The founding act, approved by Congress (the Communication Satellite Act), defined the mission as follows: to organize this technological innovation and exploit it commercially. Following an original formula proposed by the FCC (Federal Communications Commission), Comsat took the form of a new kind of private company, assuring an organic link between the state and large telecommunications companies. Half its shares were offered to the public and the other half to 163 approved firms in industry and communication. Four giants—American Telephone & Telegraph (AT&T), International Telephone and Telegraph (ITT), General Telephone & Electronics (GTE), and the Radio Corporation of America (RCA)—together held 45 percent, leaving the other 5 percent to other firms. Some 175,000 subscribers divided up the other half. On the board of directors, three representatives of the White House sat alongside the stockholders' delegates.

Armed with this operational tool, the United States proposed to the Western countries in 1964 to lay the foundations of an international network of communication by satellite. Thus originated Intelsat (International Telecommunications Satellite Consortium), with Comsat as its administrator. The hold of the United States on this system was therefore absolute. Not only did it control management via Comsat, but it owned more than 60 percent of the shares in the consortium. Great Britain, France, and Germany held a total of 20 percent, and not a single country of the Third World was among the 19 controlling nations. This American supremacy was also manifested in the area of supply contracts, with major U.S. corporations taking the lion's share. Between 1965 and 1968, scarcely a fifth of the contracts were landed by European or Japanese firms. They often limited themselves to copying American accomplishments.[39] It was not until the 1980s that American industry would be successfully countered by the European aerospace industry. This was also the decade of the deregulation of Intelsat, a public system forced, at the instigation of the Reagan administration, to face competition from private satellite systems.

In 1965, the first geostationary satellite for commercial telecommunications, Early Bird, was put into orbit, inaugurating the first generation of the international network of Intelsat satellites. Its capacity was 240 telephonic circuits or one television channel. At that date there existed on the entire planet only four ground stations capable of receiving the relayed signals, located in the United States, Great Britain, and France. The same year, the Soviet Union launched its own system of international scope (Intercosmos), conceived as the socialist camp's answer to Intelsat, and in 1971 created Intersputnik, a commercial body in which ten countries participated by the end of the decade, as against roughly a hundred in the rival system.

In signing the act creating Comsat, Congress had recommended that it "direct care and attention toward providing such services to economically less developed countries and areas"—a task it began to perform during the second half of the 1960s. A new research path in international communication opened up: prospecting the uses of satellites. Among the first nations targeted as experimental terrains were Brazil and India. At the forefront of the effort were researchers such as Wilbur Schramm, author of the first report published by UNESCO on the application of satellites to educational ends, and research institutes such as the one at Stanford.[40] It was a time when American mass communication research was in the good graces of major U.N. organizations.

Civil Reconversion

The space race as a grand narrative of the American nation-state was to last a little over ten years. The epic began to fizzle out toward 1972. "The point of view of the pragmatist has superseded that of the patriot and the pure scientist. . . . The Cold War turned temperate and a more peaceful era converged with the growing awareness of mothholes in the social fabric down here on earth." It was in these terms that a general administrator of NASA traced in April of that year the new lines of the space policy of the United States.[41]

Communication satellites, meteorological observation satellites, satellites to aid air and maritime navigation, and satellites to observe natural resources took over from moon exploration and garnered the lion's share of government budgets. Détente favored the flowering of joint Soviet-U.S. space projects (such as Soyuz). Civil reconversion of high technologies of space was proceeding apace.

Technological optimism prevailed in the major electronic and aerospace firms that sought to diversify in order to escape the monoproduction of defense equipment. With the end of the wars in Southeast Asia, the acceleration of the civil application of these technologies, whose exclusively military cycle was over, was confirmed day after day. All hopes of resolving the great social imbalances seemed warranted. As a high official of General Electric declared in 1975: "Private business should play its part in dealing with the problems of housing, education, traffic, health, waste disposal, pollution control. . . . We may see the emergence of a 'social-industrial complex,' a partnership of business and government addressed to the resolution of these major social problems."[42]

In the strategy of vertical and horizontal integration followed by the great multimedia conglomerates in the late 1960s, the sector of education and pedagogy won a place in the sun, so strong was the belief in an expanding market in this area. It was an era when psychologists, sociologists, and educators worked with television producers to find an alternative to the commercial logics of the large networks, with generous support from educational foundations such as Ford and Carnegie. The resulting programs would be transmitted by the public service network also created in the late 1960s.[43] The Black and Hispanic minorities placed on the agenda the fight against school inequalities and the integration of children from the ghettos.

The state's discourse on the use of space for the benefit of human beings in a world order where the welfare state remained a major protagonist was the theme of this "new space era." "It is necessary to offer the spirit of science and North American technology in order to resolve the problems of development. The unprecedented progress of science and technology, the demographic explosion, the explosion of communications and knowledge, demand new forms of international collaboration," declared President Nixon in 1971.[44]

"Candides" Abroad

Here again, we may ask whether there were, in the 1950s, any frameworks of research on "international communication" other than those inspired by the logic of the Cold War. Was there another way of rethinking this subject, based on experiences acquired in the war period?

The answer is yes, notably in the thought of Edward T. Hall, even if the common sense distilled by the most prominent research marginalized his way of seeing the Other beyond one's national borders. As a U.S. officer in a regiment stationed in Europe and then in the Philippines and composed essentially of black soldiers, he observed his men's difficult contact with local populations. In the 1950s, Hall worked successively in the Pacific as an intermediary between the military and the indigenous populations, and then in the Foreign Service Institute at the State Department, whose mission was to familiarize employees and diplomats with the cultures of the countries to which they were assigned.[45] Based on these overseas contacts, Hall wrote first a book, *The Silent Language*, published in 1959, followed the next year by an article in the prestigious *Harvard Business Review*, where he extrapolated conclusions for the use of businessmen.[46] His article began as follows: "With few exceptions, Americans are relative newcomers on the international business scene. Today, as in Mark Twain's time, we are all too often 'innocents abroad,' in an era when naiveté and blundering in foreign business dealings may have serious political repercussions." His conclusion was that "our present knowledge is meager, and much more research is needed before the businessman of the future can go abroad fully equipped for his work. Not only will he need to be well versed in the economics, law, and politics of the area, but he will have to understand, if not speak, the silent languages of other cultures."[47]

In this work, Hall, a representative of what came to be known as "the invisible school" or the Palo Alto school, analyzed the codes of intercultural communication. Laying the basis of "proxemics," he spoke of "cultural shocks" provoked by contacts among businessmen of different cultures. "Culture" was defined as the set of codes governing all interaction. He thus suggested that businessmen of his country become aware of the differences that governed silent languages such as those of space, time, things, material possessions, friendship patterns, and even modes of negotiating contracts or agreements. These informal languages meant, for example, that a delay in replying to a message would not be interpreted in the same way in the Near East as in the United States, because the sense of time was not the same. Similar differences of symbolic meaning underlay rules of organizing space, as witnessed by the architecture of offices, or the surface area or the number of floors occupied by the chief officer of a corporation.

This new way of seeing relations with other cultures implied a questioning of the mathematical theory of information born in the context of the "machine" universe of World War II. This theory had been formulated in 1949 by Claude Shannon, researcher in the laboratories of the Bell Telephone Company, subsidiary of AT&T, and was rapidly consecrated as the master reference of the social sciences. What the Palo Alto school contested was the legitimacy of transposing this schema—which had grown out of an attempt to make telephone communication as efficient as possible—to the field of communication among humans. Against this linear model of communication between a sender who codes and transmits a message, defined as an abstract statistical magnitude, toward a receiver who decodes it, the Palo Alto school proposed a model of circular or retroactive communication, that is, a permanent social process at several levels and in multiple contexts, integrating not just two or several "variables" but multiple modes of behavior. These are the multiple languages and codes of which Hall spoke.

This type of approach to the complexity of social communication would not be legitimized until the 1980s, a decade that witnessed not only the multiplication of intercultural relations via the market but also and above all the beginning of the United States' decline from absolute economic hegemony. The United States discovered that its ubiquitous and unshared power exercised over more than three decades had coincided with a profound ignorance of others. A study of 1,500 managers worldwide undertaken in 1989 by Columbia University showed that American executives suffered from "insularity" and "parochialism." When they were asked to list in order of importance the traits of the manager of the twenty-first century, no more than 35 percent of the Americans mentioned in first place some experience outside the country in which the headquarters of their firm was located, as against three-quarters of their counterparts in Europe, Japan, and Latin America. Asked about the impact of international events, only 17 percent of the Americans judged as "substantial" the effects of the project for a single European market in 1993, as against 34 percent of the Latin Americans and 52 percent of the Japanese.[48] But that is another story.

In the next chapter we shall take up once more the subject of war. On the terrain of military operations, Western armies had to face, in the 1950s and 1960s, another shock of cultures, and not just of ideologies: the wars of national liberation.

Chapter 5
The School of Ruse

Indirect Action

Since its beginnings, research on psychological warfare has always rendered tribute to what it considers its precursor: the Chinese Sun Tzu and his *Art of Warfare*. This ancient art, twenty-five centuries old, was defined by Raymond Aron as "a school of the ruse, of trickery, and indirect action,"[1] in a classic work where he shows both the points of rupture and the continuities between this Asian tradition of war and that other branch of strategic thought inspired by Clausewitz.

But the encounter with the rules of secular wisdom of the oldest empire in the world did not realize its full dimension until what Gilles Deleuze and Félix Guattari called, at the end of the 1970s, the "war of minorities":[2] guerilla wars, wars of national liberation, popular wars, or revolutionary wars—all these terms designated a similar reality throughout the world and yet had very different geneses.[3] This new mode of warfare appeared in the 1950s in a context marked by the nuclear threat, the indirect confrontation between superpowers and the great movement of decolonization. This war of minorities who counterposed "the movement" to the "regular state" may also be understood as a metaphor for a different model of communication: this conflict can be seen as that of a complex network of relations against a system of command both centralized and vertical, the deterritorialization of strategic space against its territorialization, nomad space against sedentary space.

It is a commonplace to note that popular war presupposes a different organic relation with the "people," the "mass," whose participation in the struggle is, by definition, an integral part of this type of conflict. The support of the population is as indispensable to the fighters as "water to a fish," as Mao Tse-tung, the foremost theoretician of popular war, put it.[4] Hence the importance given to the knowledge and analysis of internal tensions—the economic, cultural, political, and social contradictions— that run through a people and that may be exploited. Which classes, which groups are likely to join cause with the regular state? How can bridges be built and communities of meaning created around motives for struggle linking heterogeneous groups? While seeking to sap the legitimizing discourses of its adversary, the movement must produce its own narratives of mobilization and legitimation, not only via the media but through its actions on the ground. These are the "semiotic stakes of the struggle," its taking of form, the conditions of its acceptability to the "masses." These were stakes that Jean-Max Noyer, analyst of strategic thought, described well in referring to the "invention of revolt":

> Using all civil and military methods, playing on the cultural and "desire" dimensions of the confrontation, multiplying the "terrains and spaces" of conflict, bringing the conflict both onto the international scene and into the very heart of the adversary's imaginary and profiting from permanent mediatization, the popular wars are defined by a combined and more or less rationalized use of terror and pity, of calculated violence and the capacity to arouse compassion and to unleash sympathy, and the processes of "affective" investments which favor these.[5]

This moment of inventiveness proper to a period of tension leading up to the taking of power would contrast later with the lack of imagination of most of these movements once the regular state had been captured by the party-state. The search for the diversities and mediations in the approach to "the people" would lose momentum with the return to a sedentary life, with the end of what Deleuze and Guattari call the "creative line of flight of nomadic space."

Deterritorialization, for the war of minorities, is not only the nomadic space for its combatants but also its projection into the world-space thanks to media relays, as the military strategists in charge of countering them had intuited during the first popular wars of the modern era. Colonel Roger Trinquier, educated in the school of colonial wars in Indochina and Algeria, wrote:

> In view of the present-day interdependence of nations, any residual grievance within a population, no matter how localized and lacking in scope, will surely be brought by determined adversaries into the framework of the great world conflict. From a localized conflict of secondary origin and importance, they will always attempt sooner or later to bring about a generalized conflict. . . . But the

rallying of oppositions and study of effective means of protection have been neglected. More exactly, when the enemy's methods and their application have been recognized, propaganda and pressures have always been powerful enough to influence a poorly informed public and to lead it systematically to refuse to study or use the same methods.[6]

The French colonel wrote this in 1961, at a time when the media's strength of diffusion was by no means on a scale with what it would become in the following decades.

Long after the end of colonial wars, French Army officers would remember what they had experienced at the time as a betrayal: the critical attitude of the metropolitan press with respect to their actions in Algeria. Thirty years later still, during the Gulf War, when uniformed experts were omnipresent in television studios, several of them would refer to this to justify the embargo on information in time of war.

At the time when Colonel Trinquier was denouncing the danger of a "public opinion badly informed," an emblematic text on the use of the media in wars of liberation was that of Frantz Fanon, a psychiatrist from Martinique who had rallied to the Algerian National Liberation Front (FLN). In "This Is the Voice of Algeria," he wrote:

> Before 1954, radio, on the normal level, [was] already apprehended as an instrument of the occupation, as a type of violent invasion on the part of the oppressor, was, in the psychopathological realm, an evil object, anxiogenic and accursed. After 1954, the radio assumed totally new meanings. Foreign technology, which has been "digested" in the context of the national struggle, had become a fighting instrument for the people and a protective organ against anxiety.[7]

It was at the end of 1956 that the Voice of Free Algeria started broadcasting from Tunis. It is also the period when the Voice of the Arabs, the radio station of Colonel Nasser in Cairo, became the symbol of Pan-Arab revolution, and it would remain so until the Six Day War in 1967.

The "War in the Crowd"

In order to resolve the enigma posed by minority war, military thinking had to listen to its adversary.

It was the French Army that took the initiative:

> At a time when most of the world was focused on traditional ideas of general war and its doctrine of retaliation, the French were in Indochina struggling with Ho Chi-Minh's inheritance from Mao

Tse-tung—the war of "national liberation," the Communist Revolution. . . . It wasn't until Nikita Khrushchev's speech in January
1961 and President J. F. Kennedy's reaction to the latter during the
Vienna Summit, that the highest echelons of the U.S. government
began to pay serious attention to these kinds of struggle.[8]

This observation by Don A. Starry, lieutenant-colonel in the U.S. Army,
was published in *Military Review*, the theoretical journal of the American military, in February 1967, at a crucial moment in the escalation of
the wars in Southeast Asia. Already, in October 1960, in the same forum,
George Kelly, associated with Henry Kissinger at the Harvard University
Center for International Affairs, had encouraged "the most qualified
American military experts" to study "all the lessons left by the French experience" in encounters with "revolutionary war."[9] This war, widely
covered in *La Revue de Défense Nationale*, was dubbed by General
Nemo, veteran of Indochina, the "war in the crowd," two years after the
defeat at Diên Biên Phu (1954), in a prescient article published in the
same French journal.[10] The texts that captured the attention of experts at
the Pentagon and their civilian advisers were those of colonels Godard
and Trinquier, two of the architects of the "pacification" and the fight
against pro-independence forces during the Algerian War (1954–62), as
well as those of Colonel Lacheroy. As soon as it appeared in 1961, *La
guerre moderne* (Modern warfare) by Colonel Trinquier found an American publisher.[11]

For Trinquier, the unprecedented character of modern warfare, which
he also called the war of subversion, resided more in the scope of its actions (political, economic, psychological, military, etc.) than in the vagueness of the "enemy":

> In modern warfare, the enemy is far more difficult to identify. No
> physical frontier separates the two camps. The line of demarcation
> between friend and foe passes through the very heart of the nation,
> through the same village, and sometimes divides the same family. It
> is a non-physical, often ideological boundary, which must however
> be expressly delineated if we want to reach the adversary and de-
> feat him.

In military academies and classic doctrines, "one factor essential to the
conduct of modern warfare is omitted—the inhabitant. . . . Control of the
masses . . . is the master weapon of this war."[12] Lacheroy went even further: "The mass is amorphous; it is to be taken. How to take it?" "By
force or brain-washing," hastened to reply this colonel, head of the "psychological action" services, whose ambition was to wage a total struggle
for the control of populations. The fight was for their "pacification," a

term the theoreticians at the Pentagon would make their own in the 1960s.

The first element in this doctrine of pacification was the systematization of political indoctrination of the population. Numerous sources attest that those who embarked with the most virulence on this path were the officers who had spent long months of captivity in the camps of the Vietminh and were subjected to "reeducation" there. Intrigued and at the same time disconcerted by the militant practice of their enemy, just as Hitler had been by the Russian revolutionaries, they tried to turn the experience around to their use. Readings of Goebbels, Hitler, Le Bon, and Chakotin became mandatory, while the army sent a number of officers back to the classroom for training in psychology and sociology. The operational organization of "psychological action" took the form of a newspaper and leaflets, companies equipped with loudspeakers, paramilitary corps (SAS, or specialized administrative sections; SAU, or urban administrative sections), and various "civic actions" such as the building of schools and dispensaries. All these initiatives and organizational structures were accompanied by measures such as the ban on the sale of radios to any persons except those with permits granted by the security forces.

The second component of the pacification approach was the transfer of populations and their rehousing in controlled villages, "compartmentalized and sanitized," in the expression of General Massu, who justified them as "a protective measure against the minority of outlaws which causes terror to reign and imposes its will on the immense majority of good citizens." British troops were the first to engage in such massive operations in their fight against the guerillas in Malaysia in the immediate postwar period. (In fact, Her Majesty's Army had earlier resorted to similar measures during the Boer War of 1899 to 1902, copying the tactic of *reconcentrados*, which had been inaugurated by the Spanish Army a few years before, against Cuban proindependence guerillas.) This territorial and administrative control led, in Algeria, to the displacement of about a million and a half to two million Algerians.

A third element was the obtaining of "action-intelligence" by specialized teams who interrogated suspects "without holding back." Trinquier devoted numerous commentaries to the problem posed by the "visibility" of warfare operations, particularly of a "police nature in a large city." He wrote:

> [These operations] take place in the very midst of the populace, almost in public, whereas formerly they occurred on a battlefield, to which only armed forces had access. Certain harsh actions can eas-

ily pass for brutalities in the eyes of a sensitive public. And it is a fact also that, in the process of extirpating the terrorist organization from their midst, the people will be manhandled, lined up, interrogated, searched. . . . Our enemies will not fail to exploit the situation for propaganda needs. . . . Under no pretext, however, can a government permit itself to become engaged in a polemic against the forces of order in this respect, a situation that can benefit only our adversaries.[13]

What this officer called "inevitable brutalities" are not considered only a means of obtaining information at all costs on clandestine enemy networks, but also a means of destroying, in each captured individual, the sense of solidarity with an organization and a collectivity.

The principal French theoreticians of "psychological action" and of "pacification" were to be ultimately dismissed after the *putsch* of Algiers of April 22, 1961. Some of them were to join the French terrorist movement to keep Algeria French, the Secret Army Organization (OAS), or find themselves on the numerous fronts that, everywhere in the world, recruited volunteers for antisubversion crusades.

The Internal Enemy

Apart from France in this generation, the British military would be the only ones in Europe to theorize counterinsurrectional war. According to sociologists Philip Schlesinger, Graham Murdock, and Philip Elliott, to whom we owe an in-depth study of British television coverage of "terrorism" and its interpretation of the actions of the armed movements in Northern Ireland, the British variant of the theory of counterinsurgency "has much in common with the French school." Its principal representative is General Sir Frank Kitson, who, after having earned his reputation in Kenya, Malaysia, and Cyprus, commanded a brigade in Ulster in the early 1970s. His book *Low Intensity Operations* (1971) is both the result of his own experience and the synthesis of teachings handed down by French and American theoreticians.[14] The study by the three British sociologists tries to show how ideological schemas inspired by the conceptions of the struggle against "subversion" impregnated not only news reporting but also fictional representations. It remains one of the rare studies that has thought through the relation between television and political violence during a so-called "national security crisis" in a major industrial country.

The strong point of the analyses of these three authors is to show how "counterinsurrectional situations" provoked new relations between media institutions and state strategies, how "normality" and "exception" combine or are telescoped into complex modes of control and pressure that the state and the wider political establishment can bring to bear on broadcasting. This point of view is clearly summed up in their ringing conclusion-manifesto:

> This book has taken issue with the prevailing orthodoxies of Right and Left. We reject the counterinsurgents' claim that television gives extensive publicity to "terrorist" views and mobilizes sympathy and support for their causes. We also reject the commonplace radical characterization of broadcasting as a largely uncritical conduit for official views. In opposition to these one-dimensional accounts we have drawn attention to the diverse ways in which television handles "terrorism" and the problems this question poses for liberal democracies. Some programs are relatively "closed" and work wholly or mainly within the terms set by the official perspective. Others, though, are more "open" and provide space for alternative and oppositional views. The extent of this diversity, however, should not be overstated. Although television is the site of continual struggle between contending perspectives on "terrorism," the contest is not an equal one. "Open" programs appear far less frequently than "closed" ones and they reach smaller audiences.[15]

Aside from France, the United Kingdom, and the United States, it is to Latin America that one must turn to find other substantial contributions to the theory of counterinsurgency. Very dependent, in their first phase, on the Pentagon doctrines of national security, the armed forces of the subcontinent progressively disengaged from this source of initial inspiration and elaborated their own thinking about the ways of approaching "total war" against an "internal enemy"—or "pure war," as Paul Virilio called it in 1975, that is, a war without an enemy except for the one defined by oneself. This was the case notably among the Argentinian military (undoubtedly the one most influenced by the "French school" in its most extreme aspect), and the Peruvian and the Chilean armed forces. And it was above all the case of the famous Brazilian war college known as the "Sorbonne," which counted among its ranks a number of theoreticians of geopolitics, such as General Golbery, whose book on the geopolitics of Brazil dates from 1955.[16]

After the overthrow of the constitutional government of João Goulart in 1964, this science of national security would come to fruition in the creation of a state that openly avowed this counterinsurgency doctrine. The way in which the generals in power in Brasilia managed their relation with the media clearly illustrates the tension that arises in modern dictatorial regimes between the norms of "psychological war" against the "internal enemy"—all citizens being, in fact, potential suspects for the national security state—and the norms of a mass commercial culture, which caters to and seeks to seduce the consumer public. Never before in the history of the media has the contradiction appeared so clearly between two concepts and regimes of information. The military did not succeed in imposing their schemas of total control of "hearts and minds" by means of propaganda as prescribed by the doctrine, and ended up relying on the mechanisms of the market. This meant that the years of dictatorship would be both the worst years of state violence Brazil would live through—and those in which her television industry took off; two decades later, the latter would be one of the most modern, if not postmodern, on the planet. This despite a particularly ferocious censorship—a troubling coincidence which I have analyzed with Michèle Mattelart at length in another book. In this work, we wrote:

> The novelty of the Brazilian military regime was this: to assure a minimum of consensus for a political project that was forced to resort to coercion and police control, state power had to call on the commercial machinery of mass culture, the product of a society in which public opinion is a recognized actor in the public sphere. A mass culture linked to the idea of representative democracy and free access to the market economy of information, culture and entertainment. State power was thus forced to count on the mechanisms of societies in which "civil society" has a recognized, active institutional role, where "discipline-blockade" measures (Foucault) do not predominate. This is the paradox introduced into the traditional model of the authoritarian state by the modern phase of the market economy and its law of free flow of commodities and symbolic goods. . . . Closing down the "theatre of psychological operations" against the "internal enemy" is no longer conceivable in a country that has entered into an accelerated phase of industrial development and internationalization of its internal market and where television plays a pioneering role in the conquest of new market frontiers.[17]

As opposed to this system, in which television is in the hands of private groups in a market economy under a dictatorship, the South Korean television apparatus remains under the firm control of a national security state. It was not until 1990 that this newly industrialized state began a process of liberalization by authorizing foreign advertising agencies to do business in the country and by opening a debate on the creation of commercial channels.

The Science of Counterinsurgency

"Populations are amorphous or indifferent; it suffices to detect, then to form, an active élite, then to introduce it into the mass like a leaven which will act at the desired moment."[18] This is, in its simplest form, the French theoreticians' vision of social differentiation of the antisubversive struggle. Although they professed faith in the virtues of "psychological action," they remained stuck in a problematic dominated by a military logic. In short, the military demanded the principal role in all the phases of the struggle.

Their American counterparts were certainly not blind to this; they raised questions about the implications of such a doctrine with respect to the relation between army and state. Thus Don A. Starry, in the article cited above, writes:

> It is significant that throughout the empires, republics, restorations, and governments that rose and fell, the army remained loyal to France. The French army avoided politics. . . . *La guerre révolutionnaire,* as they came to understand it, challenged the very core of the profession. . . . This is a method inimical to the structure of liberal democracy. Armed forces can surely fight against the effect of subversion, but military force in a democracy is not the proper agent to deal with the causes of subversion.[19]

The Pentagon, in concert with the White House, therefore drew on the arsenal of institutions and traditions of the American state to avoid a fatal slide of this type.

Two factors favored negotiation: the existence of an institutional framework of cooperation between political power and the military, as codified by a key law, the National Security Act, and an ample supply of researchers disposed to pursue the objectives of a counterinsurgency struggle.

In 1961, President John Kennedy and his secretary of defense, Robert McNamara, undertook to modify the spirit of the Pentagon and to restructure the institution from top to bottom. The works of Ernesto (Che) Guevara and Mao Tse-tung became required reading. This was true as well for the personnel of the State Department, and more particularly those assigned to the Third World. But it was not until the first failures in Vietnam, in 1964, that the Kennedy doctrine triumphed. The failures were due precisely to overreliance of the experts on conventional war (at that time, the American intervention in Vietnam was confined to sending military advisers).

Yet in 1962, university specialists had tried to impress upon the generals the strategic importance of the study of "insurgent behavior." That year, two professors at the Massachusetts Institute of Technology (MIT), Ithiel de Sola Pool—one of the major figures in research on international communications—and Lucian Pye, had proposed, in a report published by the Smithsonian Institution and entitled *Social Science Research and National Security*, a multidisciplinary approach with a view to the "formulation of analytical models of social sciences and social control in underdeveloped countries."[20] But the military kept the report in a desk drawer. Only the failures of the programs of massive population transfers and regroupment into "strategic hamlets," as well as the growing antigovernment resistance of the Buddhists, made an alliance between the social sciences and counterinsurgency necessary.

In late 1964, the secretaries of defense and the army took stock of the resources available for this type of research and made a frank evaluation of existing studies. Their conclusion: "Primarily, there is very incomplete knowledge and understanding in depth of the internal cultural, economic, and political conditions that generate conflict between national groups."[21]

A vast program of sociological and anthropological investigation tried to fill the void by means of a research network composed of universities and private centers and financed by new bodies of ad hoc defense coordination. The research contracts in the social sciences were granted as a matter of priority to the study of minorities and elites, their values, their vulnerability to psychological operations, their social relations, and their institutions of communication. Thus much importance was attributed to the analysis of ethnic and religious minorities and the evaluation of local military authorities' ability to assume leadership in national development. The first formal models of social system engineering began to treat counterinsurgency with their input/output grid. Scenarios were invented that "simulated" the possible behaviors and reactions of "actors in the system" in insurrectional or preinsurrectional situations: the government, the military, political parties, middle classes, peasants, workers, minorities, landowners, students, business owners, multinational companies, and so on—in Southeast Asia, but also everywhere the flame of revolt was smoldering.[22]

Funds for research on counterinsurgency irrigated diverse centers of higher education: the American University, Stanford, MIT, Johns Hopkins—which included, from the end of World War II on, a research center specializing in psychological war—and many others. Ithiel de Sola Pool, who had brought himself to the attention of the Kennedy administration by developing in 1960 the first model for simulating an electoral campaign—that of the future president of the United States—lent his support to the development of a counterguerilla model dubbed "Agile-Coin." The same university professor who in the 1980s would take a strong stand in favor of deregulation would also be the only one among his colleagues, in the decisive years 1965–69, to belong to the Defense Science Board, which was responsible for supervising and evaluating the results of the alliance between research and the "national security" apparatus.[23]

The American university community was far from being unanimously in favor of participation in this kind of research, even if many of the researchers did not turn their noses up at the rewards it offered. Making an overview of the entry of defense-commissioned projects into university research since the opening of the Cold War, an editor of the review *Scientific American* did not hesitate to write in 1965:

> The new science of "human engineering" at the "man-machine interface" has brought psychology into the circle of disciplines favored by the project contract/grant. Regional research institutes, organized at the primary initiative of the Department of State and the great private foundations to illuminate hitherto dark regions on the world map, have brought sociology and anthropology into the ambience of the Department of Defense. . . . With funds abounding for projects in every field of learning, the university campus has come to harbor a new kind of *condottieri,* mercenaries of science and scholarship loaded with doctorates and ready for hire on studies done to contract specification.[24]

Quite ironically, the only input the new social engineers forgot to incorporate into their simulations was the increasing differentiation among the political actors within the United States itself. As a disillusioned Samuel P. Huntington, a political scientist who was energetic in putting his expertise at the service of counterinsurgency in Southeast Asia, would state in 1975:

> The Vietnam War and, to a lesser degree, racial issues divided elite groups as well as the mass public. In addition, the number and variety of groups whose support might be necessary had increased tremendously by the 1960s. . . . The most notable new source of national power in 1970, as compared to 1950, was the national media, meaning here the national TV networks, the national news magazines, and the major newspapers with national reach such as the *Washington Post* and the *New York Times.* There is, for instance, considerable evidence to suggest that the development of television journalism contributed to the undermining of governmental authority. . . . Television news, in short, functions as a "dis-

patriating" agency—one which portrays the conditions in society as undesirable and as getting worse.[25]

It was on the basis of this realization that Huntington, in collaboration with French sociologist Michel Crozier and Japanese sociologist Joji Watanuki, diagnosed in 1975 for the Trilateral Commission the "ungovernability of the great Western democracies."

Disinformation Under Fire

"It is my generation who must halt, then turn back the incursions which the military have made in our *civilian* system. These incursions have subverted or muffled civilian voices within the Executive branch, weakened the constitutional role and responsibility of the Congress, and laid an economic and psychological burden on the public that could be disastrous."[26] It is with these words that Senator J. William Fulbright concluded his essay *The Pentagon Propaganda Machine,* published in 1971. Subjecting to minute analysis the disproportionate development of communication strategies by the military establishment, he denounced what he called the hold of the "Defense Department's public relations activities on the shaping of public opinion."

In March 1972, Fulbright presided over a Senate committee inquiring into the activities of the USIA (United States Information Agency). Officials of the agency, diplomats, and senators threw a spotlight on the shady operations of this central propaganda body of the American government, which employed ten thousand people spread over a hundred countries. In 1966, the same senator from Arkansas had presided over the first inquiry into the USIA's lobbying of the press, in an effort to improve the image of the government.[27]

In a memorandum issued in January 1963 to the director of the USIA, President Kennedy had clarified the missions of the agency, founded in 1953, in these terms:

The mission of the U.S. Information Agency is to help achieve U.S. foreign policy objectives by (a) influencing public attitudes in other nations, and (b) advising the President, his representatives abroad, and the various departments and agencies, on the implications of foreign opinion for present and contemplated U.S. policies, programs and official statements. The influencing of attitudes is to be carried out by the overt use of the various techniques of communication—personal contact, radio broadcasting, libraries, book publication and distribution, press, motion pictures, television, exhibits, English language instruction and others.

By these means, he added, the USIA was "to give the U.S.A. the image of a strong, democratic and dynamic nation, qualified for leading the efforts made by the world in order to fulfill this purpose."[28]

In 1970, the intensification of the American involvement in Vietnam led to the assigning of a new mission to the USIA in addition to the one originally ratified by Congress: "to lend appropriate support in psychological warfare to the military command in the theater or theaters of active military operations, and provide daily guidance and basic information materials."[29] The USIA had in fact already assumed this mission on the ground by creating the Joint United States Public Affairs Office (JUS-PAO) in Saigon, in collaboration with the high command. The objective was defined peremptorily by its own officials: "winning the hearts and minds of the Vietnamese people to support the efforts of the American war, by influencing journalists favorably, learning the tactics of psychological war used by the enemy and weakening its moral strength." It was in accordance with this objective that the agency had taken part in the late 1960s in the pacification project known as Phoenix, which resulted in the "neutralization" of 20,000 opponents of the Thieu regime.

But in the course of the public hearings presided over by Senator Fulbright in 1972, it was not so much this aspect of the USIA's work that the investigation committee insisted upon. Three sources of worry were manifested in the questions addressed by its members to the agents of the propaganda body who took the stand. First, the clandestine character of certain operations that USIA agents had conducted incognito, which was expressly forbidden, since Kennedy's instructions spoke of "overt" use of the media as opposed to the methods of bodies such as the CIA, whose explicit legal mission was covert action. The second worry was the collaboration between a public agency and big American corporations in mounting certain joint operations. Finally, there was the confusion, which only grew deeper in the course of the testimony, between "propaganda" and "information." This reached the point where Fulbright had to admit that "these distinctions between information and propaganda are extremely difficult to make. They are not only theoretically difficult to make, but it is also practically difficult to draw the line."[30]

Four years later, in April 1976, almost one year to the day after the fall of Saigon, another Senate committee presided over by Frank Church summoned the heads of the military and civil secret services. Under fire were the covert actions undertaken by the CIA and the Pentagon, involving activities as diverse as propaganda, economic warfare, direct preventive action (sabotage, antisabotage), and the subversion of hostile states (assistance to guerillas or opposition movements).

Propaganda of this sort was what the veterans of World War II would have labeled "black propaganda," that is, "a fundamental intelligence operation, not only because it uses intelligence material solely as its ammunition, but because it is an independent maneuver conducted in an atmosphere of surreptitiousness. Black propaganda never identifies its real source. It pretends to originate within or close to enemy or enemy-occupied territory."[31] Under this technical definition were included the fabrication and dissemination of information and rumor.

On this precise point, the Church Commission tried above all to evaluate the dangers to which the American citizen was subjected by these clandestine practices, which the violence of the ideological clash of the Cold War had, as its protagonists admitted, "rendered indispensable despite their often frankly illegal character." It stated: "In examining the CIA's past and present use of the U.S. media, the Committee finds two reasons for concern. The first is the potential, inherent in covert media operations, for manipulating or misleading the American public. The second is the damage to the credibility and independence of a free press which may be caused by covert relationships with U.S. journalists and media organizations."[32] The most urgent grievances concerned the press campaigns orchestrated in 1970 by the CIA with the goal of destabilizing Salvador Allende, the socialist president-elect of Chile, in the two months preceding the transfer of power. Information disseminated in the framework of this covert action had been largely picked up by the *New York Times* and the *Washington Post*, thus "polluting"—this is the term used by the senator—news and commentaries. The inventory of this type of action did not stop there, since in the course of testimony from information agents the cases of Iran, Italy, Greece, Guatemala, and other countries were exposed. All these countries had in the previous two decades constituted "zones of subversion," real or potential.

These two series of hearings had the great merit of attempting to formulate the problem, and indeed of recognizing that there could be a problem regarding the exercise of democratic rights. This willingness to engage in revelations in the 1970s contrasted with the chilly attitude of other major Western democracies, where unavowed actions were classified "confidential defense" and kept in the dark. The virtue of the Fulbright and Church committees was even greater in that on the other side of the Iron Curtain, absolute opaqueness was the rule.

Dezinformatsia

The revelations of former heads of secret services who defected to the West have allowed us to lift a corner of the veil covering the international apparatus of propaganda and disinformation (*dezinformatsia*) of the former Soviet Union and its satellites.[33] This was the case, for example, when a member of the Czech politburo, General Jan Senya (one of the first high-ranking officers in this field to defect) revealed in 1968 how the secret services of the East had succeeded in 1962 in deceiving the journalists of *Der Spiegel* by leaking to them the "secret plans on NATO strategy."

In the world of Western intelligence, the prevailing opinion was that Soviet intelligence operations were conducted on a scale far greater than those of the other camp. The war of evaluations of the adversary's potential was itself an integral part of the combat of propagandists. Thus, in the Church Committee's report, it is not at all surprising to find the following estimate: the activities of the East were equivalent to six times those of the West, and even more if one counted covert operations.[34] On the front of propaganda narrowly defined, the witnesses to the Fulbright Committee subscribed to the figures that alleged a net handicap for the West: in 1970 Radio Moscow broadcast to Africa an average of 235 hours per week, in 15 languages, while the Voice of America logged no more than 130 hours per week in 4 languages only (English, French, Arabic, and Swahili). But in reply to a question from Senator Fulbright about the efficiency of such a potential, the director of Voice of America admitted that "their effectiveness is no way near commensurate with the size of their effort, because in many countries their programs are judged to be so propagandistic, so unbalanced, so strident that our information would indicate that they are not listened to other than by the party or by people who are considering joining the party."[35]

On the domestic front, the Glavlit and the "black cabinet" confiscated at the borders magazines, academic works, correspondence between scientists, and even publications of specialized branches of the United Nations, in which the Soviet delegates proclaimed themselves to be ardent defenders of science in the service of peace and understanding among peoples. One of the most convincing testimonies is an ethnographic study performed in the 1960s by Zhores Medvedev, a scientist who was interned in a psychiatric asylum for having conducted a study on the abuses of Lysenko, and who would later seek refuge in the West. But before doing so, he minutely reconstructed, with the means at hand, the itinerary of his own international correspondence and that of his foreign correspondents. Harassment, bizarre disappearances, confiscations, returns to sender, mysterious stamps on letters received, indicating their passage through indiscreet hands—all these violations of the right to individual correspondence provoked this scientist to bring suit, in what would have been a total and foregone loss were it not for the gain in knowledge about the censorship system, against the postal administration of his country, in the name of a legislation claiming to be democratic. Comparing the delivery time of letters in various parts of Europe with that at the end of the last century, Medvedev noted a singular increase in the time it took a letter from England to reach a Moscow address: it went from 3 to 4 days in 1875–1914 to 9 to 21 days in the late 1960s. At this rate, Lenin's letters sent from France or Switzerland would have taken from 10 to 17 days in 1969, while at the beginning of the century they reached his family in 3 or 4 days, despite the fact that he was pursued by the czarist police![36]

The bulk of studies on international communications performed in socialist countries in the 1970s stigmatizes the "means of bourgeois propaganda," the standard term of party rhetoric to characterize the Western media. A representative work is that of Georgi Arbatov, director of the Institute of the United States of the Academy of Sciences and a leading official in the Party's section on ideological work, whose title alone summarizes the concerns of the era: *The War of Ideas in Contemporary International Relations: The Imperialist Doctrine, Methods and Organization of Foreign Propaganda* (1970).[37] Seeing manipulation and plots wherever ideas and opinions were expressed, studies like this turned their backs on Marx's analysis of ideology as the lived experience of a way of life, transformed into the "nature of things." This approach also neglected, of course, the essential difference between the two ideological systems and their ways of envisaging the role of the media. This was a difference that the historian Roy A. Medvedev, brother of the author of the study on correspondence, explained very simply to his American colleagues: "The United States has many political parties, organizations, and religious groups; their propaganda is simply a part of the general flow of information. But in the Soviet Union, information is merely a part of the party's propaganda campaigns."[38]

With the crumbling of the communist regimes, this project of indoctrination, divorced from people's concrete daily realities, revealed its impotence, despite its massive volume, to attain the totalization of individual and collective life that had been attributed to it.

As for the U.S. government's civilian propaganda apparatus, its crisis was to occur in April 1991, although signs of it had been visible for fifteen years. For it was in that month that President George Bush, following the end of the Cold War, the democratic revolution in the East, and the events of the Gulf, would decide to commission studies on the reorganization of official radio and television so that these media might remain "competitive." This had not been the case during the U.S. Marines' intervention in Panama in December 1989, or during the Gulf War of 1990–91. These two conflicts would signal the irresistible rise of the private channel CNN (Cable News Network) as a necessary intermediary for Bush's muscled diplomacy. On these two occasions, one could observe how the pressure of the emotion of images was modifying the decision-making processes of political leaders. The cold logic of the experts was short-circuited by the immediacy of live news. In this strategic revision there was, however, an exception to the rule: the United States' policy with respect to Castro's regime in Cuba, for it continues to maintain a radio and television service in Florida (Martí) according to the old schema inherited from the Cold War.

The questioning of the objectives and methods of the USIA occurred at roughly the same time as the calling onto the carpet of the CIA, already in trouble over Irangate, the scandal involving the channeling to the Nicaraguan Contras of funds from the secret sale of arms to Iran. Although it had some 18,000 paid officials and a budget of 3.5 billion dollars, the CIA was found wanting during the invasion of Kuwait by the troops of Saddam Hussein—nor had it been able to forecast the precise moment of the fall of the Berlin Wall in 1989 or the attempted coup d'état in the Soviet Union in 1991. A number of factors contributed to the crisis of the CIA: too much emphasis on electronic espionage methods to the detriment of the human element in intelligence gathering; the difficulty in mastering and interpreting the flow of information delivered by technically sophisticated equipment; the lack of agents specialized in languages and foreign cultures; and, above all, the difficulty of ridding itself of a Manichean and conspiratorial conception of the world inherited from the struggle against "the Evil Empire" and facing up to an environment where not only the idea of the enemy becomes diffuse, but where facts about the industrial and technological strategies of rival powers become as indispensable for national security as the traditional raw material of intelligence focused on political espionage.

To cling to a one-sided definition of information is inexcusable in an age when it is becoming multifaceted. Private enterprise, beginning with the Japanese some time ago, reacted to the globalization of the economy by engaging in "competitive intelligence" or even "competitive strategy," implying that this operation concerns not only technology but all the technico-economic and political data that may figure into economic actors' definitions of their strategy. The mission of these operations was to collect, from the most diverse sources and networks (colloquia, publications, scientific and technical reports and exchanges, etc.), the "strategic information" needed by corporations in order to maintain their level of competitiveness, and to analyze it from the perspectives of various disciplines and fields of activity, calling on specialists coming from different cultures.

As for the Pentagon and its information apparatus, it was to undergo a different evolution in the wake of its future military victories.

The Gulf: A Postmodern War?

The Gulf War threw a spotlight on what had tended, in the 1980s, with the crisis of military-industrial complexes and the official celebrations of the end of the Cold War, to be a shadowy zone in the debates on the future of global systems of communication and information: the logic of surveillance and the logic of what the American geopolitical analyst James Der Derian calls the "culture of national security."[39] Who can deny, after that conflict, that the new architecture of global networks continues despite everything to play a role in this area, too?

The Gulf War was a "communications war" in two senses.

It was so first of all because of the place occupied by the censorship of information through a system of small press groups, called pools, under strict military control. The military authorized a total of eleven pools of from five to eighteen people, each covering a specific unit. These pools traveled in the war zone accompanied by a "public affairs officer" who chose the troops to be interviewed, briefed the journalists, controlled filming, and then reviewed material, suppressing any sensitive information. The result was then sent to Dhahran and distributed to the press. Access to the pools proved to be unequal, as attested by two lawsuits, one brought by American magazines such as *Harper's*, *Mother Jones*, the *Nation*, and the *Village Voice*, and the other by the *Agence France-Presse*.

From the beginning of the Gulf crisis, journalists not belonging to major media enterprises or not explicitly endorsed by governments of the coalition met with difficulty in obtaining visas from Saudi authorities. This was revealed in an inquiry conducted by the Gannett Foundation; about half the American press reporters were affected. It thus took a month for the correspondent from a Spanish-speaking channel of the United States, Univisión, to obtain his visa, while the correspondents of the major English-speaking press agencies were granted theirs in less than a week.[40]

The pool system was officially legitimated in the name of the Vietnam War precedent. The interpretation according to which television is the agent of defeat—which, as we have seen, was also that of Samuel Huntington and his colleagues—was brilliantly contested by Daniel C. Hallin.[41] In his study, published in 1986, this political scientist from the University of California showed how the Vietnam conflict had not been a "living room war," delivered live every day to the eyes of the world, as one had been led to believe. Even in the absence of government censorship, the networks were far from having shown the true horror of the war, guided as they were by self-censorship dictated as much by a certain type of relation with their audiences as by the government and the army. In other words, television had been much more a follower than an opinion leader. Such conclusions led Hallin to assert, on the occasion of the Gulf War, that "one of the most persistent myths about Vietnam is the idea that saturation coverage on television turned the public against the war, and by extension that any televised war will lose public support. This assumption motivated the British to place heavy restrictions on television during the Falklands War (1982), and is one of the major reasons the American media now face military censorship for the first time since Korea."[42]

In its classic aspect, the war of information is the heir to techniques of psychological warfare developed in the course of preceding wars. For example, the calls to desertion made to Iraqi troops had a *déjà vu* quality. To be convinced of this one need only reread the story told by David Hertz, Hollywood scriptwriter in civilian life and OSS officer in the U.S. First Army during the siege of Lorient, a submarine base in Brittany, in August 1944. This siege was a microcosm of the war in which every stratagem and ruse was used to sap the morale of some 28,000 German soldiers barricaded in their fortress.[43]

Also classic were the operations of disinformation, the production of false news or rumors, about the enemy army's potential, their losses, the spread of an oil slick, and so on—all forms of secret action about which Senators Church and Fulbright had expressed prescient disquiet, foreseeing the risk of contaminating the media and press agencies with so-called "black propaganda." One of the novelties of the Gulf War's psychological dimension was its sheer extent, made possible by the complicity of a number of journalists, either ingenuous or cynical. Nor should we neglect the pressure of the "paroxystic forms" that mass national symbols assume in time of war. In the 1950s these caused Elias Canetti to remark that such periods of blindness bring back in force the "national religion."[44]

But the qualitative change in the practices of "psychological warriors" resides in the way they molded themselves to the imaginary of mass culture. They were, of course, powerfully aided by the "new aesthetics of arms," which means that arms manufacturers are more and more attentive to the creative forms circulating in postindustrial society. It is as if, in the very design of the killing machine projected onto the TV screen, were now incorporated its dimension of media exhibition, its "desiring" dimension, its "communicating" value. The old polemic initiated by Walter Benjamin after World War I against the "aesthetic of war," dear to his compatriot Ernst Jünger, deserves to be reread.

This brings us to the second reason why the Gulf War can be defined as a "war of communication," a war of the so-called "3CI" (control, command, communication, intelligence). The army's putting into images of the aerial war made one representation spring out: the triumph of "intelligent weapons"—strategic missiles piloted by their own on-board computers; reconnaissance spy-satellites that familiarize pilots with their mission sites even before they climb into their airplanes (five satellites flew over Iraq and dispatched in real time the facts collected onto the consoles of Pentagon analysts, providing images of details as small as ten centimeters in size); systems of commands relayed to all the war apparatus and even to the weapons themselves. In short, we saw deployed before our eyes a complex expert system, "neural networks," video screen networks, and numerous computer systems, ranging from large IBMs to lightweight portables, serving either as decision centers, intermediaries for data analysis, or simply as relay stations conveying information to other systems.[45]

Moreover, by means of an impressive logistics of administration—of every hundred American soldiers present in the operation "Desert Shield," only 55 were combatants—this war of communication and of information technologies was the first to manage the "tense flows." To do so, the U.S. military borrowed the methods of management refined by Japanese auto manufacturers (computerized management, maximum diminution of stocks, ordering parts necessary for production only as required). The U.S. Air Force thus accomplished 95 percent of the logistics operations without human intervention. The major difference from the management of an auto-manufacturing corporation is that the latter administers daily no more than 100,000 references (or types of parts), while the U.S. armed forces had to manage several million. This deployment of a model of organization in time of war is such that it allows us to suppose, in the words of journalist François Came, that "the methods of management developed for the Gulf will probably change the organization of work in tomorrow's industrial world,"[46] just as the science of organization had crossed a decisive threshold in resolving the logistic problems posed by the landings on the beaches of Normandy on June 6, 1944.

Since the Falklands/Malvinas War in 1982, which pitted the army of the Argentinian dictatorship against British elite forces, the pioneering and full-scale use of means of automatic and automated destruction had raised new questions about what already appeared as an upsetting of the premises of classic strategy. Paul Virilio—in the course of his reflections on the "highest stage of speed," and the strategy of "global vision" thanks to spy-satellites, drones, and other video missiles, and above all to the appearance of a new type of headquarters—even spoke of moving beyond geopolitics and geostrategy to pure logistics, which, he says, is on the way to "becoming 'global' not only because of the range of the new 'transhorizon' weapons and the wide radius of action of missiles, but especially the temporal regime of all recent arms systems: the passage to the act of war is now only . . . the transfer of human and political responsibility from the stage of free decision of use on the battlefield to the stage of industrial and economic programming."[47]

Jérôme Binde went even further in the newspaper *Le Monde*:

> In experimental form, we come to witness, as passive voyeurs and slaves, the first purely technocratic war. . . . The actual forces are now those of computer science and communication, of the automation of the war machine. The style of this war signals its nature: neither a liberation struggle nor a national popular war, it pits against each other military "elites" of unequal strength. . . . The Falklands War is the postmodern moment where military "value" is reduced to its zero degree. . . . The technology that allows the aggressor to massacre an enemy at a distance, sparing itself the terrible face-to-face encounter, places the aggression of war at an infantile or primitive stage: that of the "omnipotence of ideas" (Freud).[48]

"Shoot and forget": this is exactly the image of war as surgical operation that the new psychological warriors have spread with images of the pilot's computer as he kills at long distance, without seeing his enemy and without being seen. This is the myth of antiseptic war between professionals.

One may be justly shocked by the proliferation of media-centered discussions on the transparency and ethics of information in time of war, and the bankruptcy of the fundamental debate on the "ethic" of this new electronic warfare itself, a veritable parade of Western technological brand names. For without such a debate, it is in our view illusory to lay the foundations of a morality for the media apparatus and its "voyeurs." This is especially true in that, in contrast to the Falklands War, the Gulf War demonstrated that the heavy publicity given to the electronic war obscured another entire dimension of the conflict. Less than two weeks after the ceasefire, the Pentagon acknowledged that the high-tech war placed in the limelight by censored news reports had represented only a small part of the action. Of the 88,500 tons of bombs dropped on Iraq and Kuwait, less than 7 percent had been laser-guided. This explains why 70 percent of the bombs missed their targets, the intelligent weapons being only used, in view of their cost, on the most precious enemy objectives.[49]

The aspiration to democracy that inspired the U.S. Senate, in the 1970s, to question the use of psychological operations in time of war is unfortunately no longer present to provoke, in that same assembly, an exposure of the communicational practices of the Gulf warriors. The consensus of the victory parades submerged any such impertinence. The moment of victory—which is also that of the death instinct—is not the time for objective evaluations.

As for the propaganda of the Baghdad dictatorship, which did not hesitate to take its own people hostage like pawns on the chessboard of a regional war, it pushed to an extreme point the crisis of the great narratives of national liberation that a large number of former colonies, now party-states, had long since transformed into hollow rhetoric, in the absence of other active social or political forces.

None of this should make us forget—even if we are careful not to reduce the Gulf War to a North/South conflict—that the logic of war has engendered simplistic thought, intolerance, and blind certitudes in media representations, not only regarding the opponents of the war within the major Western societies but also and especially regarding the rest of the world. The triumphalism of the major countries in the coalition engendered a royal disregard of long-term factors, such as the growing risk of resentment and the deepening of the gulf, at the level of culture and the imaginary, between the excluded populations and those who are integrated.

In the following chapters we shall leave the paths of war and take to the "paths of progress," which thinkers such as Proudhon, at a time when the international appeared still as the invention solely of states, could only envisage in relation to war. "International relations," he wrote in 1860, "have war as their condition and as their effect. But this manifestation of the principle that might makes right is subject to the law of progress because progress brings societies gradually to substitute the peaceful right to work for the warlike right based on force."

Part II

Progress

Chapter 6

From Progress to Communication: Conceptual Metamorphoses

The "Global Village"

What society, what world was heralded by the advent of electronic information and communication? Since the end of World War II, this question has given rise to numerous hypotheses among social scientists of various disciplines, whether researchers or government advisers. The result is that theory has been enriched by a multitude of terms and neologisms that have tried to account for the current and future changes in the status—social, economic, and cultural—of these technologies. Examining the different interests that have presided over the production and use of these concepts, theories, and doctrines and drawing up their genealogy allows us to understand what has been and continues to be at issue in these upheavals in the modes of thinking about communication. Such upheavals are marked by either abrupt ruptures or progressive displacements in the meaning of "communication," which has gone from a definition limited to the media to one with totalizing aspirations, from confinement in one industrial sector to promotion as the linchpin of a new society. The end result has been the displacement of the "ideology of progress" by the "ideology of communication."

In this genesis, one man became in himself a paradigm: the Canadian Marshall McLuhan, who wrote in 1974:

> At instant speeds the audience becomes actor, and the spectators become participants. On spaceship Earth or in the global theater the audience and the crew become actors, producers rather than consumers. They seek to program events rather than to watch them. As in so many other instances, these "effects" appear before their "causes." At instant speeds the cause and effect are at least si-

125

multaneous, and it is this dimension which naturally suggests, to all those who are accustomed to it, the need to anticipate events hopefully rather than to participate in them fatalistically. The possibility of public participation becomes a sort of technologial imperative which has been called "Lapp's Law": "If it can be done, it's got to be done"—a kind of siren wail of the evolutionary appetite.[1]

McLuhan had developed this point of view six years earlier, in collaboration with Quentin Fiore, in a book entitled *War and Peace in the Global Village*.

The Vietnam War was then at its height. The whole world, the authors asserted, was experiencing the "first television war," which meant "the end of the dichotomy between civilian and military. The public is now participant in every phase of the war, and the main actions of the war are now being fought in the American home itself." It sufficed "merely to ride the surface of change like a surfboarder." This "participation in depth" in the new environment that acted in lasting fashion on the sensorium, according to the two authors, explained why "all the non-industrial areas like China, India, and Africa are speeding ahead by means of electric technology."[2]

In this vision of the "planetary village," everything occurred by sole virtue of the technological imperative. From that point to the denial of the complexity of cultures and societies in which messages "landed" and produced effects was only one step, which other analysts, immersed in the war of ideas, did not hesitate to take. Seizing on this determinist conception, they read into it what they had already been convinced of for a long time: the new technologies of communication meant the end of ideologies and the rise of a new idea of social change that rendered totally obsolete the old obsession with political revolutions. For according to these followers of McLuhan and Fiore, the "communications revolution" had already begun to resolve the problems that political revolutions had never come close to resolving.

Thus the Venezuelan writer Arturo Uslar Pietri did not hesitate to cross the Rubicon in 1972 by asserting:

> The world revolution has already begun in the United States. . . .
> The unprecedented political fact of the rejection by a great part of
> the people of the United States of the war in Vietnam, or the situa-
> tion of the black minority, was not the fruit of any one particular
> political group, or of any precise ideology, or of an abrupt change
> in the management of public affairs. Fundamentally, it was the re-
> sult of the volume, the intensity, and the range of the media. . . .

The man in the street is first of all transformed into an eyewitness and then inevitably participates in these events which would have remained faraway and insignificant in other circumstances.

He concluded with this prophecy: "We will all succeed in participating in everything that happens and we will react accordingly, outside the bounds of ideology, models, and commands."[3]

In 1972 the broad American public was tuned into the slogan of the "communications revolution." This new leitmotif took off during the second half of the 1960s. To ensure its circulation among a mass audience, marketing and advertising agents did not bother with nuances, proclaiming from the outset: "The communications revolution which has taken place over the last seven years has stimulated the desire to consume, collective social responsiblity, youth revolt, the revolt of women, fashion revolt and the era of individual judgment: in short, a new society."[4]

Throughout the world, the "communications revolution" would of course have its popularizers and technicians, but it would also have its writers and ideologues, avowed anticommunists, who in their best-selling books would turn the "technological revolution" into a new warhorse in their struggle against everything on their left, classifying those who were not in agreement with this new redemptive myth in the authoritarian camp.[5]

The Anarchist Crucible

Because they were turned into frozen forms in the market of "instant thinking," in which ideas of global apocalypse or catharsis circulate at an ever faster pace, McLuhan's analyses, over time, lost the mark of their origins, as is well demonstrated by James W. Carey in his study of the genealogy of modern media theory.[6] It was forgotten that they belonged to a precise intellectual tradition and that they had evolved a great deal since the Canadian author published his first book in 1951.[7]

His first writings left the door open to an examination of the potential for tightening of social control inherent in the new electronic technology. The more his work progressed, the more technological determinism took the upper hand and became an unbridled optimism, and the more his critique of "industrialism" eroded. More and more McLuhan also left behind the analysis of the vast cultural complexes in which technologies evolve and from which they derive their meaning. In addition, McLuhan's thought cannot be explained without reference to its affiliations with contemporary or past authors, who inspired him, as he openly acknowledged, without his necessarily espousing all their ideas. The two crucial influences are the Canadian Harold Innis and the American Lewis Mumford. The latter, in turn, cannot be understood without reference to two other figures, the Russian geographer and anarchist Peter Kropotkin (1842–1921) and the Scot Patrick Geddes (1854–1932).

The analyses of Mumford, Geddes, and Kropotkin had begun in the 1920s to revolutionize the way of looking at the relation between city and country, and had opened the way to another concept of urban planning. All three had in common the same central concern and the same intuition: communications technologies are "extensions of man" and technological change is at the center of the history of civilization. Mumford, from his first work, *Technics and Civilization* (1934), borrowed from Geddes the distinction between the "paleotechnic" (steam-driven and mechanical technologies) and the "neotechnic" (electricity).[8] He also later adopted Kropotkin's utopia, prophesying that electricity would provide the way out of the machine era and restore the life of communities. Harold Innis was a sociologist, who, unlike the eclectic and publicity-minded McLuhan, devoted himself to laying the foundations of a theory not only of the effects of technologies on society but also of the cause of innovations and changes in communications, with much reference to geography and economics.[9] In *The Gutenberg Galaxy* (1962), doubtless his most important book, McLuhan goes so far as to say: "Harold Innis was the first person to hit upon the process of change as implicit in the forms of media technology. The present book is a footnote of explanation to his work."[10] This led one commentator on McLuhan's work, Thomas W. Cooper, to remark that the famous slogan "The medium is the message" is a "daring distillation of Innis."[11]

The thought of these writers, as James W. Carey notes, went in different directions: while McLuhan little by little lost the critical spirit of the first years and turned into a prophet of a new age of grace, Innis never lost sight of his project of renewing the political economy of the media; Mumford became more and more distant from his initial utopianism in which technical networks of communication were endowed with redemptive virtues. Above and beyond these thinkers' different evolutions, one must remember that their major impact is to have exploded the postulate of the priority of content over form, that is, to have insisted on the fact that the medium itself determines the character of what is communicated and leads to a new type of civilization. This postulate, nurtured by pedagogues of Enlightenment philosophy and the "typographic men" of the Gutenberg galaxy, had for too long inhibited understanding of the nature of changes brought about by electronic networks.

A Radiant Future

"Reliance on television—and hence the tendency to replace language with imagery which is international rather than national, and to include war coverage or scenes of hunger in places as distant as, for example, India—creates a somewhat more cosmopolitan, though highly impressionistic, involvement in global affairs."[12] These words of Zbigniew Brzezinski, dating from 1969, show not only that one could conceive the problem of "participation" in a very different way but also that the enigma of the nature of the new cultural environment posed by McLuhan, which he thought he had resolved with a metaphor, was taxing a number of minds.

It was in fact in the 1960s that the first academic statements were made on the social, economic, and political nature of the technological mutation in communications and, consequently, on its new international dimension. Among the pioneers were Columbia University sociologist Daniel Bell and political scientist Zbigniew Brzezinski—also a Columbia professor, an expert on the problems of communism, and future national security adviser under President Jimmy Carter. To Daniel Bell is attributed the concept of "postindustrial society." Brzezinski is credited with that of "technetronic society." At the center of their concerns was how to anticipate and prepare for the future of the society that arose from the Industrial Revolution.

What these studies and many others dating from the 1960s and 1970s clearly had in common was their effort to escape from an exclusively media-oriented perspective, as it had been fashioned in the sociology of communication and mass culture, and to place the media in a larger context of a new technological system of "communications."

While Bell's book on postindustrial society dates from 1973, his thinking on the subject goes back further in time.[13] It in fact began in 1959 with his first contributions to the Salzburg seminar. But it appears to be the Commission of the Year 2000 of the American Academy of Arts and Sciences, established in the United States in 1965 and chaired by Bell himself, that brought these analyses into vogue. This commission had been created to pursue the work of futurologist Herman Kahn, a physicist by training.[14] The term "postindustrial" is not Bell's, however. It goes back to the 1920s and its inventor was the British socialist Arthur J. Penty, author of two books, *Old Worlds for New: A Study of the Post-Industrial State* and *Postindustrialism.*[15]

It was in an article published in 1968 that Bell identified in the postindustrial society five "dimensions" that distinguished it from the previous society: 1) the creation of a service economy (dominance of the tertiary sector); 2) the predominance of a class of specialists and technicians; 3) the importance of theoretical knowledge as the source of innovation and change (the development of products of new industries originates in work realized in the pure sciences, through what Bell called the "codification of theoretical knowledge"); 4) the creation of a "new technology of intelligence" with its panoply of tools such as systems analysis, games theory, decision theory, and so on; 5) the possibility of autonomous technological growth.[16]

Formulated in the 1960s, this definition presumed that preponderant tendencies registered in the course of these years of prosperity in the great industrial nations would continue under conditions of sustained, exponential growth. The vision is thus attributable in large part to the optimism of the era. It springs from a linear conception of history, shared as much by W. W. Rostow's theory of modernization—to which we will return in the next chapter—as by the perspectives of Kahn's Hudson Institute. The initial hypotheses of Bell that have most suffered from the social and economic evolutions of the 1970s are, first, those referring to the continuation of the "affluent society"; the unshakeable confidence in a science and technology both neutral and autonomous as the motor of social and industrial innovation, controlled in the final analysis by an elite of specialists; and finally, the everlasting character of the welfare state and the state as provider of funds to leading sectors of industry, in particular the "military-industrial complex." The first hypothesis would soon be called into question by the energy crisis of 1973. As for the second, history was to remind us that everything that is technically feasible is not economically viable or socially acceptable. The state's tendency to retreat and leave the field open to market mechanisms would strike the third hypothesis a stinging blow in the early 1980s. Along with it collapsed the idea of a class of state scientific advisers, planning society's future from on high thanks to their theoretical wisdom.

Bell asserted that postindustrial society would stress two types of services, the "social services" (such as education, health, social security) and "professional services" (computing, systems analysis, and scientific research). One thing is clear: the strong growth tendency of these "social services," which he observed in societies governed by notions of the welfare state and public service, has been widely supplanted, because of the crisis of these notions, by the spectacular development of "professional services." In addition, the swelling of the tertiary sector by parasitic occupations and by services where arms and legs count more than intellectual functions, would tend to relativize the myth of the imminent advent of a service society based on the rising force of science and technology alone. "Tendency is not destiny," said Lewis Mumford—but this idea had escaped the first generation of theoreticians of the postindustrial society in their search for an "ideal type" of future society.

The Fetish of Science

Bell's thinking is contemporaneous with another perspective, to the formalization of which it strongly contributed: that of the "end of ideology," which complements his theory of postindustrial society.

In a book published in 1962, Bell had in fact settled his accounts with the question of ideology. "The most important latent function of ideology," he wrote, "is to tap emotion. . . . Ideology fuses these energies [of masses] and channels them into politics."[17] Like a number of political scientists such as Seymour Martin Lipset and Edward Shils, he arrived at the conclusion that in Western society, ideology, defined as the realm of emotions and of passion stimulated by class war, would no longer have a role to play, since the fundamental political problems posed by the organization of society and of industrial democracy had been resolved.[18] Rationality reigned in societies of abundance, which defined themselves as stable. This was the reason they were on the road to a higher stage of postindustrial democracy and economy. The mission of Western societies was to stimulate the development of politically and economically free institutions, in their image and likeness, throughout the rest of the world.

The troublemakers in the scheme were the "intellectuals" who, unlike the "scientists" acting on the basis of "objectivity," were fueled by subjectivity or ideology. The intellectual, wrote Bell, pushes into the forefront "*his* experience, *his* individual perceptions of the world, *his* privileges and deprivations, and judges the world by these sensibilities."[19] To move the social sciences out of this subjectivism, Bell speaks of the necessity of circumscribing with precision the "ideological and scientific components" of social theory. That is the price one must pay for attaining objectivity. By this measure, the ideologue is necessarily "someone else," who is labeled a propagandist. The debate is therefore not really one at all, since the dice are loaded. They have been so since the first political use of the word "ideologue" in a disparaging sense by Napoléon, who, by deforming the word forged in 1796 by the philosopher Destutt de Tracy, thus designated the liberals who opposed his regime. Destutt de Tracy had created the term "ideologist" to characterize the political and philosophical group of which he was part. With Napoléon, ideology became a synonym for "hollow analysis or discussion of abstract ideas."[20]

Many of the points in Bell's diagnosis were not as original as they appeared. They had already provided the frame for the argument of Raymond Aron's book *L'opium des intellectuels* (The opium of the intellectuals), published in 1955.[21] Ideology is for Aron the contemporary expression of old millenarianisms. The chief exponents of this new form of millenarianism are seen by Aron as veritable religious leaders.

But at that time, the majority of the intellectual class of Europe was on the side of those who thought that the thesis of the death of ideologies was itself just an ideology. And those political scientists in the United States who did not refrain from expressing their misgivings with respect to Lipset, Bell, and his colleagues understood this well when one of them, J. La Palombara, wrote in 1966:

> [It was as if] ideology had necessarily to be effaced by the great
> leap in public education, the means of mass communication, the
> multiplication of dishwashers, cars and televisions. . . . One may
> say, without risk, that one could find thousands of European intel-
> lectuals, and tens of millions of other Europeans, who would react
> to this assertion with irony or total suspicion. . . . The assumption
> of this thesis is that the future will consist of national histories
> which will be the monotonous replications of "anglo-American"
> history. In short, they would have us believe that "they" are get-
> ting more like "us" every day.[22]

Europe in this period proved in fact that there were other readings of the phenomenon of "ideology." Empiricism seemed to European intellectuals to be yet another version of the fetish of science. In the 1950s Roland Barthes had, in his *Mythologies*, demystified the claim that this particular conception of the world and of society was situated outside ideology. "Bourgeois ideology," he wrote,

> is of scientist or the intuitive kind, it records facts and perceives
> values, but refuses explanations; the order of the world can be seen
> as sufficient or ineffable, it is never seen as significant. Finally, the
> basic idea of a perfectible mobile world produces the inverted
> image of an unchanging humanity, characterized by an indefinite
> repetition of identity. In a word, in the contemporary bourgeois so-
> ciety, the passage from the real to the ideological is defined as that
> from an *anti-physis* to a *pseudo-physis*.[23]

The notion of myth appeared to Barthes to account for this passage from the real to the ideological. Myth voids social phenonema of their reality, deprives them of their history, and domesticates them by integrating them into "the nature of things." In doing so, myth renders innocent and purifies the existing order, and passes off its subjectivity and singularity as parameters of objectivity and universality—the law of some as the law of all. It can easily be seen that this approach to the ideological is scarcely compatible with that other one, consecrated not only by Bell but by the whole of empiricist sociology. The contrast is demonstrated by this definition taken from the American *Dictionary of Social Sciences* in its 1964 edition: "Ideology is a pattern of beliefs and concepts (both factual and normative) which purport to explain complex social phenomena with a view to directing and simplifying socio-political choices facing individuals and groups."[24] We see better, too, why there was such a misunderstanding regarding the myth of the end of ideologies, deeply rooted in the conservative tradition, and why this myth had such a long life. It recurs systematically in every period in which there is an attempt to silence all those who believe that the world is perfectible and that the invariable cannot be the premise of all variations. Is there any better proof that ideology does exist and can surely not be defined narrowly as "a simplifying model to direct choices"? To be convinced of its vitality, one need only invoke the latest of its avatars, the myth of the "end of history," recycled at the end of 1989 by Francis Fukuyama, strategy expert at the U.S. State Department. The fact that transistor radios have become a common item in China, that Mozart serves as background music in Japanese supermarkets, and that rock music in Prague was an expression of revolt against a moribund Stalinist ideology was for this 36-year-old neoconservative an irrefutable sign of the democratic homogenization of the world under the sign of Western capitalism.[25] As the cover of *New Republic* retorted: "Is History over? Tell it to Peru."

The argument of subjectivism used by Bell and his colleagues to discredit critical sociology in the name of science is a clone of the argument brandished by the bureaucrats of the Eastern countries against dissident intellectuals who, despite harassment and the threat of the gulag, continued to do their job of being thinkers of society. And the mystical belief in science and technology was at least as powerful among establishment sociologists such as Bell, Rostow, and Kahn, with their theory of scientific and technical revolution, as among their Eastern European colleagues of the *nomenklatura* who elevated science to the status of a "direct productive force."

The "Global City"

With Brzezinski, we enter into the geopolitics of empire in the age of scientific and technological revolution. In contrast to Bell, whom he recognizes as a precursor, Brzezinski in his book *Between Two Ages*, subtitled *America's Role in the Technetronic Era*, examines world-space and the space occupied by the U.S. superpower in its competition with the other superpower, the Soviet Union. As the title makes plain, the United States was going through a transition period. It was the first country to leave the industrial era and enter the "technetronic era," a neologism he preferred to Bell's expression because it seemed to him to express better and more directly the principal forces behind the change. Brzezinski stresses that, just as one did not refer to the industrial society that followed agricultural society as a "postagricultural society," it is preferable to designate the new era and the new society with a term proper to it. The "technetronic society" is a "society that is shaped culturally, psychologically, and economically by the impact of technology and electronics—particularly in the area of computers and communications."[26] The technological characteristics he brings out do not differ substantially from those noted by Bell, but his analysis privileges the new social and cultural complexity. At the root of this complexity is the global character of political processes. In the era of classic international politics, the four factors of power and integration were essentially national in scope: weapons, means of communication, economy, and ideology. These four factors are now becoming worldwide in character.

The notion of globalness is therefore central. Its "obvious, immediate cause" is communications. Means of communication and computers had created "an extraordinarily interwoven society" whose paradox is that reality (but also humanity) is growing unified and fragmented at the same time. "While our immediate reality is being fragmented," writes Brzezinski, "global reality increasingly absorbs the individual, involves him, and even occasionally overwhelms him."[27]

Brzezinski ventured to predict, however, that the new reality would not be that of the "global village," because "McLuhan's striking analogy overlooks the personal stability, interpersonal intimacy, implicitly shared values, and traditions that were important ingredients of the primitive village."[28] The analogy of the "global city" seemed to him to respond better to the technetronic society: "global city" because it is "a nervous, agitated, tense, and fragmented web of interdependent relations." Rather than speaking of relations of intimacy, one should speak of reciprocal influence, that is, "interdependence," another key concept for Brzezinski. It is not his invention, but with him it enters into the operational references of the diplomacy of international economy. By 1945, the word "interdependence" had made its reappearance; the political scientist Christian Richard was to suggest a biomorphic analogy. It was to be conceived as a solidarity of an organic type, resting on the differentiation of functions and the division of labor as they exist among organs of a living body. By introducing this term into the language of political philosophy, Richard thought that it would be henceforth impossible to establish a durable peace while conserving for each nation the right to absolute sovereignty.[29] Brzezinski, for his part, began to think that this political level of meaning was not sufficient and that the technetronic revolution demanded the reordering of international economic relations. With the legitimation of a new world division of labor came the legitimation of the freedom of action of multinational corporations, those great economic units whose expansion does not easily harmonize with the defense of the idea of sovereignty.

If Brzezinski refused to adhere to McLuhan, it was precisely because the interdependence within the "global nervous system" is afflicted with "occasional malfunctions because of blackouts or breakdowns," unlike "the mutual confidence and reciprocally reinforcing stability that are characteristic of village intimacy."[30] As for the notion of "homogeneity" in the modern world of tomorrow, nothing could be less certain. On the one hand, it is strongly threatened by "intellectual fragmentation," that is, the growing gap between the pace of expansion of knowledge and that of its assimilation. This "raises a perplexing question concerning the prospects for mankind's intellectual unity." What must be avoided at all costs is that the homogeneity be that "of insecurity, of uncertainty, and of intellectual anarchy"—all factors of instability.[31]

"The first global society in history" was the United States. It is "the principal global disseminator of the technetronic revolution." It is the society that "communicates" more than any other, since 65 percent of total world communications originate there, and it is furthest ahead in the development of a world information grid. But above all, it is the only country to have succeeded in proposing a global model of modernity, of patterns of behavior and universal values. Precisely because of the global character of U.S. society, according to Brzezinski, it is increasingly inadequate to speak of its world influence and its relation with other peoples in terms of "imperialism." This term is useful, he believes, only for the brief period of a "transitory and rather spontaneous response to the vaccuum created by World War II and to the subsequent felt threat from communism." The spread of the technological and scientific revolution "made in the U.S.A." has radically changed the terms of the problem. The strength of this revolution is such that it "compels imitation of the more advanced by the less advanced and stimulates the export of new techniques, methods, and organizational skills from the former to the latter."[32]

Entropy and Transparency

In the history of the concepts and conceptions that have molded thought on the role of communicating machines in social organization, a scientist had the first intuition of the structuring character of the new technology: Norbert Wiener, the father of cybernetics. In his work *Cybernetics: Con-*

trol and Communication in the Animal and Machine (1948), written when the world was scarcely at the dawn of computer science, he diagnosed that future society would be organized around "information." At this time was constituted the stock of arguments that would be used in the following decades by the extreme partisans of computerization as well as by its adversaries.[33]

There is nothing abnormal about this ambivalence. Wiener in fact presented the novel ideal of an "information society" while putting us on guard against the risks of its perversion. The main enemy was entropy, that is, the tendency of nature to destroy what is ordered and to precipitate biological degradation and social disorder. As Wiener wrote in his introduction: "The notion of the amount of information attaches itself very naturally to a classical notion in statistical mechanics: that of *entropy*. Just as the amount of information in a system is a measure of its degree of organization, so the entropy of a system is a measure of its degree of disorganization; and the one is simply the negative of the other."[34] Information, the machines that process it, and the networks they weave are alone in the fight against entropy. The information society can only be a society where information circulates without impediment. It is by definition incompatible with embargo or secrecy, inequality of access, and the transformation of everything that moves in channels of communication into commodities. The persistence of these factors can only favor the advance of entropy—in other words, push back progress.

Although he was abstract in his exposition of the laws of a modern theory of communication, Wiener became hyperconcrete when it came to designating the obstacles to a free circulation of information. Thus, in the final chapter of the original edition of his book, Wiener was implacable in his analysis of the institutions of power. At the same time, he showed us what he understood by the links between "information, language, and society," the title of the concluding chapter. "Of all the anti-homeostatic factors in society," he wrote,

> the control of means of communication is the most efficient and the most important. One of the lessons of the present book is that any organism is held together in this action by the possession of means for the acquisition, use, retention, and transmission of information. In a society too large for the direct contact of its members, these means are the press, both as it concerns books and as it concerns newspapers, the radio, the telephone system, the telegraph, the posts, the theater, the movies, the schools, and the church. . . . Thus on all sides we have a triple constriction of the means of communication: the elimination of the less profitable means in favor of the more profitable; the fact that these means are

in the hands of the very limited class of wealthy men, and thus naturally express the opinion of that class; and the further fact that, as one of the chief avenues to political and personal power, they attract above all those ambitious for such power. That system which more than all others should contribute to social homeostasis is thrown directly into the hands of those most concerned in the game of power and money.[35]

Transparency, the refusal of social exclusion, and questioning the logic of the market: here in any case are three issues present in Wiener but absent from the thinkers of postindustrial or technetronic society.

The Test of the Balance of Payments

From a more specialized perspective, a small number of economists tried, during the 1960s and early 1970s, to give a more precise content to the generic term "information."

One of the more serious efforts is the one undertaken by a specialist in the evaluation of the balance of payments, Fritz Machlup, an economics professor at New York University and formerly at Princeton. His book, which appeared in 1962[36] and was still a standard work of reference 30 years later, studies the production and distribution of knowledge in the United States, an economic activity that weighs more and more significantly in the national budget. Although all information in the ordinary sense of the word is knowledge, according to Machlup, all knowledge cannot be called information. His analytic schema distinguished five types of knowledge: 1) practical knowledge, useful in a person's work, decisions, and actions (professional, business, or trade-related knowledge, as well as political and household knowledge); 2) intellectual knowledge, which satisfies intellectual curiosity, as part of liberal education, scientific training, or general culture, and which requires "active concentration"; 3) small-talk and "pastime" knowledge, which satisfies the desire for light entertainment and emotional stimulation (local gossip, jokes, news of crimes and accidents, light novels, stories, games) or anything that demands only a passive attitude; 4) spiritual knowledge, related to religion; and 5) unwanted knowledge, accidentally acquired and not long retained. All these categories constitute the basis of what he calls "the knowledge industry."

In the discussion of his criteria of classification, one point particularly revealing of the change then in progress is the very definition of culture, which was more and more bound up with the apparatus of mass culture. Also revealed was the unease it created in an elderly economist in quest of a parameter. Machlup admitted his bias—and foresaw the objections of "snobbery" or "elitism" that would be leveled at him—in separating "entertainment" from high culture. He refused to rally to the official terminology that lumps together, under the heading of "entertainment," films as well as opera and concerts as well as variety shows. He also thought he had resolved a dilemma already faced by one of his colleagues, Anthony Downs: not wanting to settle the question once and for all, Downs had seen fit to weld the two terms together with the term *entertainment knowledge* and thus to amalgamate the two notions.[37] This same prejudice prompted Machlup to classify advertising under the heading "unwanted knowledge"!

The other entry point chosen by Machlup in dividing the knowledge industry into different categories was the occupational approach. Drawing up an inventory of different types of "communicators" or "producers of knowledge," he enumerates six: 1) the transporter, who delivers a message without changing it; 2) the transformer, who changes the form of the message, as the stenographer does; 3) the processor, who changes form and content, but only by following routine procedures (rearrangement, combinations, calculations), such as an accountant preparing statements; 4) the interpreter, who changes form and content by using his or her imagination, as, for example, a translator; 5) the analyzer, who uses her or his own judgment and intuition in addition to accepted procedures in such a way that the transmitted message will have little or no resemblance to the one received; and 6) the original creator, who, from a store of received information from messages of all kinds, adds his or her inventive genius and creative imagination in such a way that there is little in common between what is received from others and what he or she then communicates.

Starting with this conceptual schema, Machlup analyzes in detail the activities and occupations of the "knowledge industry" and evaluates their contributions to the national product: "education" as "knowledge acquisition" (from formal education, or education in private or public school, to training on the job, instruction in church or training in the armed services); "research and development" (basic research and applied R and D); the "media of communication" (printed matter, photography and phonography, cinema, stage, broadcasting, advertising and public relations, postal service, telegraph and telephone, conventions); "information machines" in use in the three sectors of education, R and D, and the communication industries (printing trade machinery, musical instruments, motion picture apparatus and equipment, telephone and telegraph equipment, signaling devices, measuring and controlling instruments, typewriters, electronic computers); "information services" offered by professionals (legal, engineering and architectural, accounting and auditing, medical); "financial services" (banking, securities brokerage, insurance, real estate); the intelligence services of wholesale traders; and government services.

Machlup notes the difficulty of circumscribing statistically all the constitutive elements of the knowledge industry. Among the insurmountable obstacles is the existence of services that are not objects of market transactions. This is particularly true when it comes to evaluating the cost of government-subsidized public education. "Most of the services of the knowledge industry," he writes, "are not sold in the market but instead distributed below cost or without charge, the cost being paid for in part or in full by government (as in the case of public school). . . . Hence we lack the valuations which for most other industries the consumer puts on the product by paying a price for it. There are no 'total sales' and no selling prices." Despite the lack of reliability of some of the available statistics, Machlup arrives at the following diagnosis: between 1940 and 1959, the work force of the United States grew by 23 percent while those occupied in the knowledge industry increased by 80 percent; the group of managers and civil servants registered the biggest increase (103 percent). As for the proportion of the knowledge industry in the gross national product, it represented, in the late 1950s, approximately 29 percent. In the course of the preceding years, its annual rate of increase (10.6 percent) had been double that of the GNP, with production in other sectors counting for 4.1 percent.

Machlup hoped that his study would serve as a basis for a reform of the educational system whose productivity he measured. He was not aiming to make a direct contribution to debates that had already made the structural leap of speaking not only of "industry" but also of the "information society" (or its numerous synonyms). Nothing in the work of this pioneer of economic measurement links him to that sort of theoretical perspective on the future of society.

On the other hand, developing such a perspective was the central concern of the report by Marc Uri Porat published in July 1977 by the U.S. government in no less than nine volumes. Its object was to define and measure the weight of the economy of information in American society.[38] After having noted more than 70 industries and over 6,000 products that could in one way or another be classified as "information," it arrived at the conclusion that in 1967, primary and secondary information—the classification depending on whether the information was exchanged on the market or not—constituted 46 percent of the GNP of the United States, with 25 percent coming from the production, processing, and distribution of goods and services of information, and 21 percent from the production of information services by private and public bureaucracies for their internal use. "Information workers" alone earned more than 53 percent of the aggregate wages. The industrial sector, which still represented around 40 percent of the economy in 1946, counted for no more than 25 percent of the work force some twenty years later. The information sector had shot up to 47 percent.

Marc Porat divided information services into three fundamental categories: 1) financial information, insurance, and accounting as well as information contained in data bases; 2) cultural information (films, television, radio, books, newspapers, magazines, news bulletins); and 3) knowledge-information (patents, management, advice). In "knowledge," he thus included all "know-how" and "show-how," organizational experience, scientific and technical information, and management information. In 1973, the export of "knowledge" represented almost ten times that of films and television programs. This led Porat to state that henceforth the problem of a country's cultural dependence would become more complex because "a technological system of production, once installed, is a most enduring cultural artefact."[39] He pointed to the implications of this new situation for the formulation of U.S. foreign policy, referring explicitly to the accusations of "cultural imperialism" that, at the time, were leveled by major international organizations.

The "Brain of the Planet"

It was in the second half of the 1970s that these perspectives started to make real inroads in Europe, among international institutions and in governments. Three anecdotes are worth recalling. First, the meeting of experts organized in Paris in 1975 by the Organization for Economic Cooperation and Development (OECD) to discuss the implications of the marriage of computer technology and telecommunications based its discussion on some documents drafted by the sociologist Ithiel de Sola Pool, Marc Porat, and one of his collaborators, Edwin Parker—all three Americans;[40] second, the first colloquium on social science research in the realm of telecommunications in April 1977, sponsored by the French National Center for Study of Telecommunications (CNET) and with the collaboration of the National Center for Scientific Research (CNRS), counted among its rare foreign guests the same Ithiel de Sola Pool;[41] and finally, Daniel Bell was one of the stars—by video link-up—of the week-long symposium on "Computers and Society" that took place in Paris in the fall of 1979.

Such events merely translated into French the effervescence already surrounding these themes. In 1978 appeared a report by Simon Nora and Alain Minc, entitled *L'informatisation de la société* (The computerization of society), commissioned two years earlier by President Giscard d'Estaing. For the first time in Europe, a major industrial country fully measured the importance of the technological change and formulated the broad lines of a national policy in this area. Only one other government had up to then produced a comparable document: Japan, in 1970, with its superministry of International Trade and Industry (MITI). It published another such report in 1979, a year that also saw the proposals of the Clyne Commission in Canada. Meanwhile, Tokyo had already drawn operational lessons and taken measures designed to encourage the expansion of its electronic firms abroad. In 1972, this ministry proposed to the Diet a law for the protection of foreign Japanese investments.

From the French report came the term *télématique* (telematics), fruit of the growing interpenetration of computer technology (in French, *informatique*) and telecommunications. To confront an economic and political crisis that they did not hesitate to qualify as civilizational, Nora and Minc proposed investing in the new information technologies. Telematics would open a "radically new horizon, because it conveys information, which is power." Because of its structuring role, information would preside over the establishment of a "new global mode of regulating society." This "new nervous system of organizations and of an entire society" would "recreate the informational agora, enlarged to the dimension of the modern nation." The new telematic networks were situated at the heart of the redefinition of relations between citizen and State, between civil society and state. The decentralizing virtues of telematics were expected to protect the fragile equilibrium between the uncontested dominion of public authority and the self-management of citizens and thus to allow the full blossoming of civil society.

This therapeutic vision of the technology of information as guarantor of the reconstruction of social consensus was accompanied by an industrial strategy for developing the telecommunications-computing tandem. The new technologies were seen as the means for a New Deal: the realization of this anticrisis strategy seemed inevitable to the authors of the report if France wished to keep her place in an increasingly global market and in the context of mounting competition. On the horizon could be seen the specter of the supremacy of IBM, the "IBM challenge." Arguing for a policy of national independence, Nora and Minc wrote:

> Data banks are often international, and the development of transmissions allows access to them without excessive tariff penalties from any point on the globe. Hence the temptation in some countries to utilize American data banks without setting up their own. Indifference to this phenomenon is based on the belief that this de-

pendence will be no stronger and no more disturbing than for any other type of supply. But the risk here is of a different character. Information is inseparable from its organization and its mode of storage. . . . Knowledge will end up by being shaped, as it always has been, by the available stock of information. Leaving to others—i.e., to American data banks—the responsibility for organizing this "collective memory," while being content to plumb it, is to accept a form of cultural alienation. Installing data banks is an imperative of national sovereignty.[42]

Two official reports in the early 1980s, more specific about the flow of data across borders, would confirm this diagnosis on the international stakes of the control of these flows.[43] "The essential stake," warned the president of the Commission française sur les flux transfrontières des données, Alain Madec, in 1980, "remains the disposition of territory on a world scale, and in particular the localization of the higher tertiary activities: 'the brain of the planet.'"[44]

The second report about knowledge in the most developed societies, written in 1979 by the philosopher Jean-François Lyotard at the request of the Council of Universities of the government of Québec and published under the title *La condition postmoderne* (The postmodern condition), also stressed that knowledge was "the major stake, maybe the most important, in the world competition for power." But Lyotard mainly examined what seemed to him an effect of the progress of science: the crisis of "certain social categories delegated to the functions of critical subject" (the Third World, parties, student youth, eponyms of the revolution, the great hero, etc.), and the parallel crumbling of what he called the metanarratives, the grand narratives of legitimation, great perils, great journeys, and the major goals (for example, the discourse of the Enlightenment, that of Hegel on the fulfillment of Spirit, or that of the Marxists on the emancipation of the workers). This crisis, he predicted, would increasingly undermine the function of nation-states:

The notion that learning falls within the purview of the state as the brain or mind of society will become more and more outdated with the increasing strength of the opposing principle, according to which society exists and progresses only if the messages circulating within it are rich in information and easy to decode. The ideology of communicational "transparence," which goes hand in hand with the commercialization of knowledge, will begin to perceive the state as a factor of opacity and "noise."[45]

Linking the change in the function of states with the rise in power of a new type of knowledge elite, the philosopher concluded:

For brevity's sake, suffice it to say that the functions of regulation, and therefore of reproduction, are and will be further withdrawn from administrators and entrusted to machines. Increasingly, the central question is becoming who will have access to the information these machines must have in storage to guarantee that the right decisions are made. Access to data is, and will continue to be, the prerogative of experts of all stripes. The ruling class is and will continue to be the decision makers.[46]

Interpreting Lyotard's hypotheses about a future composed only of small narratives of legitimation as a determinist exhortation to distrust the slightest inclination to concerted social action, psychoanalysts such as Félix Guattari did not fail to see in this diagnosis of the postmodern condition the "paradigm of all submission, all compromise with the existing status quo."[47]

Only isolated thinkers such as Jacques Ellul dissented from the idea that electronic technologies would play the role of social prosthesis.[48] For some time already, Ellul's ideas had traveled beyond the limits of the French arena and begun to influence American thought on the ethics of technical progress. Since the early 1950s, this Protestant thinker suggested that technology, which was formerly only a "means," had become a "milieu." Only by taking into account the social function that new "thinking machines" are destined to play was it possible in his view to understand the implications of technology as mediator. Profoundly skeptical about the democratic potential of these new forms of purely technical integration, he has never stopped denouncing the severe social imbalances that this "milieu," which had become a veritable technicist system, was causing for citizens. For Ellul, who did not shrink from speaking of the "technological bluff," the wired-up society is a new form of social management, in which what predominates are collections of individuals with no interaction other than that created by and through technology.

Vulnerable Societies

Defense specialists added a new term to the national sovereignty equation put forward by the Nora-Minc Report: Western security. "Having lost control of energy sources and primary materials," wrote the former French ambassador to NATO in 1978, "are we now vulnerable and threatened? . . . One countermeasure should doubtless be that the West devote itself more and more to activities that best use the most grey matter and the highest technicity. . . . If it is possible for developing countries to acquire our technologies, it is almost exclusively in the West that these are elaborated and refined. Should we envisage a nontransfer of our advantages if the resources necessary to us are refused us?"[49]

Security, national security, vulnerability: these terms would haunt the discourse of numerous states on the role of information technologies. This was true in the superpowers as well as in smaller states not tempted by hegemonic ambitions.

In 1980 Sweden produced a very controversial report bearing the title *The Vulnerability of the Computerized Society*. Its central idea was that the economy is in danger of breaking down because computer systems are vulnerable to acts of war, terrorism, economic embargo, and to error. The authors of the report recommended that the government conceive legislation that would provide the country with an effective mechanism for prevention and for action in case of crisis. They proposed drawing up a list of "K-enterprises," that is, the companies that were most dependent on electronic information and therefore most likely to become targets. They even argued that in case of an "information breakdown," only 15 percent of the national potential for production and distribution would continue to function.[50]

In 1983 the U.S. Senate issued its report on the broad lines of a foreign policy in the domain of information and telecommunications.[51] This official document led to the 1982 Amendment to the Communications Act, an important reform of the basic law on the U.S. communications system, originally passed in 1934. The Senate had been assigned the mission of drawing up an inventory of the problems that the new hegemony of "information" as a raw material risked engendering for the "political and economic security" of the United States. This measure was passed in the context of growing protectionist attitudes expressed by countries in the large international organizations in the name of national sovereignty. The report took the opposite position, resting on the doctrine that only the principles of the free flow of information and competition on the open market could guarantee the protection of everyone's interests, beginning with those of the United States.

In any event, it was not until 1991, at the time of the Gulf War, that the major industrial countries discovered the real complexity of the relation between security and transfer of knowledge. The fear of renewed military confrontation with a regional power led them to postulate the necessity of the "nonproliferation" of "destabilizing technologies." It was therefore necessary to reduce the export of so-called double-use technologies, that is, those that could be diverted by the importing country to bring its own arsenal up to date.[52] The countries of the South quickly saw in this proposition a way of stopping their development in the area of high technologies.[53] Was it not through the initial transfer of technology that large countries such as India and Brazil were able to develop their own computing and space industries or to produce fiber optics?[54] It is also true, of course, that the appropriation of foreign technologies has propelled the Latin American giant into the club of the six foremost arms exporters in the world.

The Abortion of Communist Universalism

What would Zbigniew Brzezinski think more than twenty years after the publication of *Between Two Ages*, having returned to his studies after Jimmy Carter's reelection failure? A question was put to him in December 1990 by Michel Foucher, a French specialist in geopolitics: Do you believe that the dominant factor of power in the 1990s will be military potential, economic success, or cultural or ideological influence? Brzezinski's reply:

> The military factor will surely lose its importance, with the disappearance of the Soviet threat. Consequently, it is possible that economic success and cultural influence will assume greater and greater importance. Which brings us back to the question of ideology. I think that today we risk believing too naively in models of wealth and market freedom of the United States, Japan, and Germany. . . . Whatever the case, the basis of American power is in large part its domination of the world market of communications. Eighty percent of the words and images that circulate in the world come from the United States.[55]

The political and economic debacle of the communist countries has changed the situation from top to bottom. It sealed the failure of the internationalist project dreamed of by Marx and Engels in their *Manifesto* as a counterpoint to the supranational vocation of capital. The utopia of internationalism conceived as the abolition of the nation did not prove strong enough to prevent the successive fissures in the solidarity between "proletarians of all countries," which the founders of the doctrine imagined as a "class without national interests." Not only did "the nation" and "nationalism" gain in strength in the countries that had earlier subscribed to internationalism, but the state and its "bureaucratic and military machine," whose disappearance Marx predicted under the reign of communism, invaded society to the point of strangling it.

Communist universalism succumbed to communist dogmatism and became a communism of sects and of mutual excommunications. In 1979, examining the "societies which secrete boredom" and their inadequate transition to modernity, Brzezinski wrote presciently:

The tragedy of communism as a universal perspective is that it came both too early and too late. It was too early to be a source of true internationalism, because mankind was only just awakening to national self-awareness and because the limited technological means of communication available were not yet ready to reinforce a universal perspective. It came too late for the industrial West, because nationalism and liberal concepts of state-reformism preempted its humanist appeal through the nation-state. It came too early for the preindustrial East, where it served as the ideological alarm clock for the dormant masses, stimulating in them increasingly radical nationalism.[56]

On the other hand, what the theoretician of the technetronic society could not foresee was the durability, in the era of the "global city," of the old "gunboat diplomacy" that he had declared obsolete, in spite of the rise of "network diplomacy." In other words, there was still a tension between geopolitics and geoeconomics, or between political logics and financial ones in the setting of the rules of the new world order.

With the perspective provided by the Gulf War and the struggle for control of world oil supplies, one could better measure the pitfalls of discourse on high-tech information as the way out of economic crisis. In such discourse, information is presented as a new primary material that is replacing the traditional energy sources as the line of demarcation between rich and poor. The cleavage between rich and poor countries has in fact never ceased to mark the history of strategies and theories of international communication, thus exposing the hidden side of progress, that is, the one left in the shadows by the ideology of redemptive technology.

Chapter 7

The Revolution of Rising Expectations

Against Fatalism: Modernization

"Progress is the development of order," wrote Auguste Comte in the *Catéchisme positiviste*. "Progress consists of nothing but the development of order," he reiterated in the *Discours sur l'ensemble du positivisme*. Some 130 years later, in 1968, Robert McNamara, president of the World Bank and former secretary of state and of defense under John Kennedy, wrote: "Security is development and without development there can be no security."[1]

The expansion of the great powers in the nineteenth century took place under the banner of "civilization." A civilizing mission was assumed by the merchant, the soldier, and the missionary. In the name of Western civilization, numerous peoples were denied the right to govern themselves. World War I had renewed the alibi of an imperial and colonial order. It appears clearly, for example, in President Woodrow Wilson's "Fourteen Points" speech delivered on January 8, 1918, and in the founding treaty of the League of Nations signed in June of 1919. In its first statements, the League spoke of "peoples not yet capable of governing themselves in the particularly difficult conditions of the modern world," adding that "the well-being and development of these peoples form a sacred mission of civilization," and suggesting, "the best method of realizing this principle is to confide the care of these people to the developed nations." The notion of "development" here remains very close to that of "civilization," and belongs more to the cultural and social domain than to the economic. It was only gradually that the slide to the latter took place. Some League of Nations texts refer timidly to economic development in the 1930s, but it would not become a truly mobilizing concept until after World War II.

It was in 1949 that the notion of "development" appeared in the language of international relations, designating by its antonym—"underdevelopment"—the state of that part of the planet that did not yet have access to the benefits of progress (one of the three pillars of the religion of humanity, dear to Comte). The expression was born in the White House and passed into history via the 1949 State of the Union speech given by President Truman, in a section entitled "Point Four." This program aimed to mobilize energies and public opinion to combat the great social disequilibria that threatened to open the door to world communism. The ideology of progress metamorphosed into the ideology of development. Communication and its technologies were called on to occupy a key position in the battle for development.

It was ten years later, in 1958, that the first book appeared on the possible contribution of the media to "development" of the "Third World": *The Passing of Traditional Society* by Daniel Lerner, a specialist in psychological warfare and author of an earlier work entitled *Sykewar: Psychological Warfare against Germany, D-Day to VE-Day* (1949). The subtitle of his book on the waning of traditional society was *Modernizing the Middle East*,[2] and the book contains survey results from six countries in the Middle East. It proposes to evaluate the exposure of different categories of the population to programs broadcast by international radio stations through 1,600 interviews performed in Turkey, Lebanon, Egypt, Syria, Jordan, and Iran. A seventh country, Iraq, was to be included, but the team of researchers had to interrupt its work there for political reasons. At the time, the region had been already classified by geopolitical analysts as strategic for reasons of "oil equilibrium." In 1951, the Iranian Prime Minister Mossadegh had nationalized the oil wells and refineries; two years later, he had been overthrown in a coup d'état carried out with the complicity of the CIA. In 1956, the nationalization of the Suez Canal by Nasser in Egypt provoked a conflict with Israel, England, and France.

The audience study used by Lerner had been commissioned in 1950 by Leo Lowenthal of the Bureau of Applied Social Research at Columbia University, directed by Paul Lazarsfeld. Lowenthal was responsible in the years 1949–54 for the "Evaluation of Radio Programs" branch of the International Broadcasting Service, which was under the auspices of the State Department. The actual purpose of this research, undertaken in 1950–1951, was to determine the position of Voice of America with respect to its direct competitors, above all the BBC and Radio Moscow. Armed with a research guide of some 50 pages, the native researchers, trained by a representative from Columbia on the spot, had interviewed some 325 people in Jordan for two to three hours. Divided into five categories (desert Bedouins, village farmers, the urban and rural bourgeoisie, and elites), the sample group had been questioned on its exposure to the media (radio, press, cinema) and on its opinions on local, national, and international affairs. Certain questions aimed to test the opinions of the sample group on the Korean War, the idea of "Soviet imperialism," "American imperialism," and the question of Palestine.[3] But beyond the study of audiences, the aim of the research was to collect the most possible material for drawing up a typology of attitudes to "development." It seemed natural to Lerner to apply mechanically what he had been taught by his research on psychological warfare as an intelligence officer in World War II. There he had discerned three categories: the "moderns" (assimilable to anti-Nazi categories) who were already converted and who best combined the three forms of mobility proper to modern society (physical, social, and psychic); the "traditionals" (identified as the Nazis in his previous studies); and the "transitionals" or "apoliticals," the category of people most likely to respond to the propaganda effort.[4] In this book as in the sectoral reports on this research, the term "Westernization" served as a password to define the modern attitude and the "cosmopolitan taste" of certain categories of audience.

Modernization = development: as soon as Daniel Lerner's book became a classic, that equation was confirmed. It would be backed up by other later books: *The Achieving Society* by David McLelland (1961); *Communications and Political Development* (1963), published under the editorship of Lucian Pye; and *Mass Media and National Development* by Wilbur Schramm (1964).[5] The latter book was even reissued by UNESCO and became a standard reference book for that organization throughout the decade. UNESCO, which began to concern itself seriously with the media after 1962, spared no effort to spread the theses of these authors by publishing in several languages an anthology of essays edited by Bert Hoselitz and Wilbert Moore under the title *Industrialization and Society*.[6] The preeminence of American sociology in international circles aroused the British sociologist Jeremy Tunstall to make the following biting commentary: "Daniel Lerner, Ithiel de Sola Pool and Wilbur Schramm in the 1960s became a sort of travelling circus . . . advising first this Asian government and then that U.S. federal agency. Daniel Lerner was the intellectual leader of the circus. Ithiel de Sola Pool was the commissar of the group—one of the U.S. Department of Defense's most vigorous academic spokesmen, and a vigorous anti-communist. . . . The third member of the circus was Schramm . . . the travelling salesman. Based at Stanford, Schramm in the 1960s became UNESCO's favourite mass media 'expert.' "[7]

Working within the tradition of individual psychology, Lerner developed the concept of empathy, which he defined as "psychic mobility" or "the ability to project oneself into the role of another." The problem was to free "traditional" people from their inertia and fatalism by getting them to adopt the mobility that defined the modern attitude. A great many sociologists and political scientists attempted to draw up the portrait of the "modern man," defined essentially by mental flexibility in the face of new situations and assimilation of the broad value orientations of Western industrial societies.[8] In their own fashion they transposed and transformed into a ready-made formula the analyses of Max Weber on the Protestant ethic and the spirit of capitalism, which saw in the "entrepreneur" a singular combination of the desire for personal achievement, innovative research, the spirit of competition, the quest for profit, and methodological honesty.[9]

Above and beyond its different variants, the theory of modernization exhibited a single matrix. Social change was defined as the *linear* passage between traditional society and modern society (industrial, urban, and Western). The negative pole concentrated within it all the handicaps: it was a society, culture, or personality that was static, homogeneous, frozen in time, governed by a single value system and by undifferentiated institutions, dominated by values beyond the individual—such as tradition, magic, divinity, ancestors, the sacred—and allergic to other cultures and little disposed to assimilate them. In short, it represented a society that completely resists cultural change. The other pole combined all the advantages for breaking free of that condition: modern society is driven by a conscious and voluntary transformation; change in such a society is an institutionalized, normal state of things, and it is required by the growing application of science and technology to all spheres of social life; values are no longer ascribed by tradition and accepted passively, but are modeled according to criteria of efficacy and rationality by an individual who is secularized and free to make choices. Modern society is open, turned outwards, and cosmopolitan; its institutions are specialized and segmented.

The first deficiency of this mode of understanding societies, which proceeds by establishing and confronting "ideal" types or "models," is clearly that of not referring to anything but itself, since it is entirely a construct. The least one can object is that it leaves hidden in the shadows the epistemological and cultural premises of those who advance it. This is why, once it was applied in the field, it would soon run up against the resistance and opacity of the real. The problem with this evolutionary conception of development is not only its tendency to generalize and thus to deprive each Third World society of its history and align realities and cultures that are very dissimilar. A further problem is its aversion, bordering on contempt, to the "traditional," which is seen as incapable of giving rise to anything else. Thus David McLelland, who like most other representatives of the theory of modernization did research on the influence of the "need for achievement" (a concept very close to that of empathy) on economic growth, saw in traditional cultures, judged as affective and irrational, only obstacles to development. For this theoretician of the "achieving society" (modern society par exellence), "managerial" culture has no use for traditional cultures except perhaps to preserve them out of mere historic interest.

A "Non-Communist Manifesto"

In this "smooth" vision of social change, the media occupy a central place: they are envisaged as agents of development, as producers of modern forms of conduct. They are bearers of the "revolution of rising expectations," stimulated by models of consumption and aspirations displayed by those who have already acceded to the high stage in human evolution represented by modern society.

This thesis is synthesized in a text by Ithiel de Sola Pool published by UNESCO in English, French, and Spanish in 1963:

> The propaganda in favor of modernism contained in commercial communications media is not solely intended to obtain sales for a particular brand of soap. It certainly aids this operation, but it would have neither audience nor effect if the communications media did not provide a product much richer in savor or excitement. Persuasion towards a particular choice is only part of a general argument for a totally modernized mode of life. The communications media, whose object is to open the market to new products and new interests, also present the image of a new kind of man in a new kind of milieu. As Marx underlined, the businessman is a revolutionary, even though this is not his intention. It is the mass media which transform what would otherwise be the unrealized dream of a few modernizers into the dynamic aspiration of a whole people.[10]

The media reflect technological and social modernity at the same time as they transmit it to elites. And both spread to the backward sectors of a country. On the one hand, the population is cut up into "reference groups" with different "opinion leaders," and on the other, it is a passive mass. Here we rediscover in action the theory of persuasion by stages, developed two decades earlier by Lazarsfeld, Katz, and their collaborators in their research on the decision-making of electors as "consumers."

The media act as "ferrymen" who bring traditional individuals across and onto the shores of "progress"—for this theory does basically rest on an idea of progress. This conception is well illustrated by economists of modernization such as Walt W. Rostow, who did not hesitate to subtitle his 1960 founding book on the stages of growth *A Non-Communist Manifesto.*[11] Beginning with a historical analysis of the industrial development of England, he infers from it his linear model of "stages of development" through which each country must pass. The model was universal: every society seeking to undertake the transition from a "traditional society"—pre-Newtonian in its conception of science and technology—to that of the "age of mass consumption" was obliged to repeat the industrialization experience of those that had preceded it. In moving gradually through these stages, it would accede to progress, measured essentially by the growth of the per capita gross domestic product. In this passage from one society and one economy to another, the so-called takeoff stage was crucial, since it was followed by development in the proper sense and then maturity, the threshold of the phase of high consumption.

This evolutionist vision is not far from the "institutional phases" that, according to Daniel Lerner, all candidates for development had to cross: "Everywhere . . . increasing urbanization has tended to raise literacy; rising literacy has tended to increase media exposure; increasing media exposure has 'gone with' wider economic participation (per capita income) and political participation (voting). The model evolved in the West is an historical fact. . . . The same basic model reappears in virtually all modernizing societies on all continents of the world, regardless of variations in race, color and creed."[12]

From this quantitative theory of development there flows an index approach, which would be seized upon by civil servants of international bodies such as UNESCO. The "modernizers" and their model of causal explanation by index had determined, for example, that only a country in which 10 percent of the population resided in cities could expect to "take off" in literacy. This minimum was also required for the takeoff of the media, which, from this threshold, could aspire to grow, in tandem with urbanization, up to 25 percent, and so on. They attempted as well to classify all the different "underdeveloped" countries according to these scales. The correlation between indices became the key. "Exposure to the media," measured according to newspapers, newsprint, radios, and cinema seats per capita, was correlated with income per capita, the literacy rate, and the rates of urbanization and industrialization. From these emanated a strategy for change based on a charter of "minimal standards." To escape from underdevelopment, a country had to possess, for each

100 inhabitants, at least 10 copies of newspapers, 5 radios, 2 television sets, and 2 cinema seats.[13]

All these indices and correlations were then assimilated by political scientists who derived from them models of "political development." Urbanization was combined with literacy, exposure to the media, and voting participation. As British author Peter Golding, one of the sharpest critics of the various versions of modernization theory in communication studies, noted:

> Peasants come to town, learn to read, study the newspapers, and vote wisely. . . . Political development is taken to be a dependent variable. Urbanization is not dependent on any other variable in the system. It is necessary to assume unidirectional causation. . . . What is political development? Indicators award points on the basis of reliance on political parties and free elections and on the number of political leaders out of office. . . . The other indices can also be criticized on the general grounds that they are devoid of any concern for content. Thus education is the number of pupils in schools, not a system of cultural transmission; mass communication is the number of radios, with no concern for their use. This weakness derives from the assumption that political development is a process of integration and social cohesion provided normatively by the existence of a communications network.[14]

The New Nation-Builders

The cascade of titles appearing on the theme of development or modernization in the first half of the 1960s reflects an intensification of the risks involved in this process. It now became impossible to defer the fight against underdevelopment, under pain of losing out to elites tempted by the model of violent revolution. At this point of no return, the question of development was posed as a problem of security. But this third term in the equation only rarely appeared in the writings of sociologists of development. In fact, it was only referred to openly by Lucian Pye, professor of political science at MIT, former intelligence officer in China, and most importantly, author of one of the first sociological analyses of modern guerillas, the movement led by communist forces against the British in Malaysia.[15]

Yet it is scarcely possible to understand the theory of modernization without this repressive side, that is, in close relation with the doctrines of counterinsurgency and national security. One reason is that in this period, during which modernization gave theoretical justification to the goal of development, it also legitimated the rise to power of the military in various countries. Between 1967 and 1972, the number of countries governed by military chiefs of staff more than doubled. It reached the point that the military caste appeared to several theoreticians of modernization as the group best suited to take up the challenge of development and of "nation-building." In fact, a number of American political scientists centered their work around this notion in the 1960s. "Nation-building," wrote one of them, "is presumably a metaphoric rubric for the social process or processes by which a national consciousness appears in certain groups and which, through a more or less institutionalized social structure, act to attain political autonomy for their society."[16] According to these sociologists, the army was cut out to be this chosen group on the level of national consciousness. Professional capability, equipment, manpower, systems of sanction, and models of emulation—all these assets made the army the best guarantor of the national project.

It was from this perspective that in 1961 Lucian Pye inaugurated, at the demand of the Pentagon, a series of studies that opened with this preamble: "There is urgent need for systematic research into potentialities of military establishments for guiding economic development and assisting in the administration of national policies."[17] In the same anthology in which Pye's remarks appeared, another sociologist, J. J. Johnson, was even more blunt:

> Only a few years ago it was generally assumed that the future of the newly emergent states would be determined largely by the activities of the Westernized intellectuals, their nationalist ruling parties, and possibly their menacing Communist parties. . . . Now that the military had become the key decision-making element in at least eight of the Afro-Asian countries, we are confronted with the awkward fact that there has been almost no scholarly research on the role of the military in the political development of new states.[18]

The most patent result of these appeals from the academic community was the preparation and publication, in the course of the decade, of an impressive series of monographs on military power in environments as different as Egypt, the Middle East, Brazil, Chile, Peru, and Indonesia.

In this stream of studies, the conceptual frameworks developed by Lucian Pye around "nation-building" were those that referred primarily to the media as the expression of the modernizing project. What is most noteworthy is his perspicacity in discerning the tensions inherent in the drive to modernize. Thus he wrote in a study on Burma: "At the heart of the problem of nation-building is the question of how the diffusion of the world culture can be facilitated while its disruptive consequences are minimized."[19]

In the key year of 1961, President Kennedy persuaded Congress to vote the Foreign Assistance Act, which authorized the administration to aid programs of "civic action" undertaken by military authorities in the Third World. The Defense Department wrote this notion into its glossary the following year, defining it as "the use of preponderantly indigenous military forces on projects useful to the local population at all levels in such fields as education, training, public works, agriculture, transportation, communications, health, sanitation, and others contributing to economic and social development, which would also serve to improve the standing of the military forces with the population."[20] In fact, however, since many armed forces were caught up in counterinsurgency struggle, few of them assumed this new Good Samaritan role. The "civic action" through which the sorcerer's apprentices of Pentagon-financed research tried to link, within legal bounds, military power and the civilian life of a nation, would later prove to constitute in many cases the first step on the path to the violent seizure of power.

The second reason why it is impossible to dissociate the modernizing strategy from the politics of security is that the heyday of the former is contemporaneous with the U.S. government's Public Safety Programs. This phenomenon was depicted in the 1972 film by Costa-Gavras, *State of Siege*, which had been inspired by an authoritarian modernization project in Uruguay. The retraining of Uruguayan police forces took place under the aegis of the Office of Public Safety—launched by the Kennedy Administration—of the United States Agency for International Development (USAID), a division of the State Department. USAID sent experts and up-to-date equipment designed to counter the "internal enemy"; it also transferred knowledge through training programs in which courses on mass psychology, the pathology of insurgents, the use of photographic techniques in demonstrations and civil disturbances, and many other subjects found in no official university curriculum mobilized the knowledge and know-how of the sociology of communication.[21]

Finally, it may be useful to remember that the military side of the strategies of modernization also manifested, near the end—that is, in the second half of the 1970s, before their total loss of legitimacy—moments of large-scale paranoia. How else can we understand the pouring of the most elaborate electronic and military aerospace technologies into the Shah's Iran and its secret police, the SAVAK? In 1974–75, the export of weapons to this country, which the United States saw as an unconditional policeman in the Gulf, represented 40 percent of the total sales of the U.S. arms industry abroad. As a unique instance in the annals of the aerospace industry, the installing of Iran's national system of satellites was not under the orders of either NASA or Comsat but the U.S. Air Force.

This was the case because, for the first and last time in a Third World country, the military uses of the system were to coexist with its civil functions, including the relaying of programs, data, and telephone conversations. U.S. strategists at the time considered it crucial to install a very important intelligence base, capable of tapping all electronic conversation in the Gulf region (project IBEX). Iran signed the "contract of the century" with an American firm for the modernization of its national telecommunications network. It was less than four years later, in 1979, that this technological megamachine under high military surveillance crumbled like a house of cards under the patient labor of the ayatollahs and their network of followers, armed with their audiocassettes.[22]

Throttling the Population Explosion

In August 1961, a brand-new program furnished the framework for an ambitious modernization project, this time a civilian one: the Alliance for Progress, an aid program for the "takeoff" of Latin American economies. A veritable Marshall Plan for development, its ambition was to create new forms of international cooperation with the countries of the hemisphere and new political formulas for accomplishing the "revolution of freedom" in order to counter the extension of Castro's revolution, then in its triumphant stage. This pacific revolution claimed to break with the old U.S. tradition of support for traditional oligarchies and military dictatorships by supporting the rising middle classes.

While this offer of collaboration took as a priority the countries of Latin America, in fact the whole of the Third World was targeted by this philosophy of modernization. Its concrete effects were felt in three areas: family planning, rural innovations, and new educational technologies. In all these domains, the "diffusion of modern attitudes" was at the forefront.

If the order of the day in the 1960s was the fight against underdevelopment, it was also the struggle against the population explosion. In the minds of development strategists, the fate of the two were clearly linked. Efforts at economic takeoff would be in vain if, at the same time, one did not curb fertility rates in the Third World. Discussions became more and more frequent within various bodies of the United Nations, as aid organizations redefined the issue and the objectives of the struggle were more clearly focused. Public opinion was mobilized. In a *New York Times* article in April 1965 under the suggestive title "We Help Build the Population Bomb," the agronomist William Vogt, veteran exponent of extreme methods, denounced what he saw as the pusillanimity of official U.S. policy: "We should provide active help in all aspects of birth control as

freely as we now help to build steel mills or hydroelectric projects in foreign countries. . . . We can continue to follow our present path, increasing population pressures in much of the underdeveloped world. This is
the way of disaster. . . . Or we can make our aid programs truly helpful
by including the one element—population balance—without which all
the other economic factors are useless."[23]

The position of the president of the United States at the time, Lyndon
Johnson, was also unambiguous. In a statement given on the twentieth
anniversary commemorative session of the U.N. General Assembly in
June 1965, he stated: "Let us act on the fact that less than five dollars invested in population control is worth a hundred dollars invested in economic growth."[24] To accelerate the implementation of this policy, Johnson proceeded completely to recast USAID's strategy in this area.[25]

The new charter that resulted, while stipulating that the organization
would abstain from dictating policies and imposing choice of method,
stated that USAID would consider all requests for cooperation concerning demographic studies, but also provide technical assistance, including
the training of specialists in family planning. The great American educational foundations contributed substantially to this redeployment by
bringing fresh funds and expertise. The Population Council, a private
body founded in 1952 and subsidized by the Ford and Rockefeller Foundations, transformed itself into a logistical support center and a center
for elaborating doctrines and strategies. The sociology of communication
and demographic sociology (still called the sociology of population) contributed jointly to this enterprise.[26]

What was expected of the social sciences was first to prepare the
ground for the adoption of this innovation by polling the target populations. This was the function of so-called AUK studies (Attitude, Use,
Knowledge). Their aim was to measure the attitudes of the population toward the introduction of birth control techniques, then the knowledge
that the target population had of such techniques, and, if applicable, the
use they made of them. The sociologists in charge of these studies made
no mystery of their philosophy of action. As J. M. Stycos of Cornell University wrote in the *Public Opinion Quarterly*:

> The main function of these investigations is similar to any market
> study: showing that a demand for the goods and services exists, in
> this case, a demand for birth control. . . . These studies represent,
> moreover, a way of beginning an action without attracting contro-
> versy. As well as supplying information useful for eventual future
> programs, the research itself stimulates the interest of people di-
> rectly or indirectly implicated and can accelerate the whole policy-
> making process.[27]

In this research, as in all projects based on the principle and theory of diffusion, and following the techniques of advertising persuasion that it copied, the means by which an idea was disseminated from its source toward its final users was rigorously codified into stages: awareness, interest, evaluation, trial, and adoption or rejection.

The second function of the sociology of mass communications in the expansion of family planning was to furnish the programs with expertise on the most effective use of the media and the best motivational schemes for initiating the process of persuasion. The sociological literature of this period displays both naiveté and cynicism in its massive adoption of modern sales techniques with a view to creating modern attitudes, particularly among women and men of the popular classes.[28] What better symbol can be imagined than this: the sociologists of the Population Council believed in those years that there was no better reward for each convert to vasectomy in India than a transistor radio! The modernization cycle had looped around on itself.[29]

Despite this deployment of massive policies of birth control, the annual report of the United Nations on its population activities estimated in 1991 that if the decision were made to control the rate of population growth by the year 2000—and that would mean not exceeding 6.4 billion inhabitants, one billion more than at the beginning of the decade—$4.5 billion would be needed.

Experiments among the Peasantry

It was thought that policies of modernization in the countryside could obviate more radical agrarian reforms, such as those involving redistribution of land. The "Green Revolution" came along at the right moment. Formally, it designated only a "genetic revolution": by hybridization, American scientists, mobilized thanks to the financial support of the Rockefeller Foundation in particular, had obtained new and much more productive varieties of grain seeds, notably rice and wheat. But in the context of ideological competition, the "Green Revolution" was quickly converted into a slogan: against the so-called political solution, it offered a technical solution to a dangerous social problem. These new varieties of seeds, combined with an injection of new agricultural technologies, new chemical fertilizers, and new methods of irrigation, were presented as the best means of eradicating underdevelopment and hunger from the world.

Several theoreticians settled down to the task and developed the theory of the spread of innovations, or "diffusionism." Neither the term itself nor the theories date in fact from this period; they were intimately associated with classical ethnology in the last decades of the nineteenth century. German diffusionism, associated with Ratzel, and the British version, associated with the triumvirate of W. H. R. Rivers, Elliot Smith, and W. J. Perry, were at the center of the debate on the "mode of diffusion of progress," on the laws of development or evolution conceived in terms of phases, on the "active" and the "passive" races, and more generally on the relations between the "developed" and the "primitive" models of civilization and culture. This debate referred back directly to the discussion of the concept of "imitation" in the work of the Frenchman Gabriel Tarde and his embryonic theory of the diffusion of cultural traits. In defining imitation as a spontaneous and irrational act that induces inferior individuals and classes to copy, or "ape," their superiors and, above all, in defining imitation as the sole determinant of the social bond, several epigones of diffusionism had already revealed the elitist and ethnocentric premises of their work.

The major representative of the more recent incarnation of diffusionism was Everett Rogers of Stanford University, whose first book, *The Diffusion of Innovations,* was published in 1962, followed by *Modernization among Peasants*, published in 1969. In the latter, Rogers perfectly synthesized the concept of development that guided his concern about modernization. He wrote: "Development is a type of social change in which new ideas are introduced into a social system in order to produce higher per-capita incomes and levels of living through more modern production methods and improved social organization."[30] One cannot help noting once more the tendency to tautology peculiar to this current of thought, which, just as in advertising, finesses the definition of the "new" or the "improved"—or rather, takes them as given.

Rogers's research covered all the fields of application of modernization, from family planning to technologies,[31] but it focused above all on the peasantry. Consequently it belongs to a whole tradition of research in rural sociology that began in the United States in the 1920s and took off in the 1940s, and which chose to observe the process of adoption of technical innovations by farmers. Rogers's work is one of the best illustrations of what happens when theories of persuasion by steps are extrapolated to contexts different from that of the United States. In fact, from these hypotheses he elaborated a typology of farmers ("innovators," "early adopters," "early majority," "late majority," and "laggards").

In the 1960s, these axes of research influenced many rural development projects from Colombia to India. But they quickly became subject to criticism coming from this same Third World, criticism that extended to all the research deriving from the theory of modernization. The most judicious and constant objections came from sociologists specialized in rural communication and rural extension.[32]

In general, what the critics of the diffusion model refuted was its supposed neutrality and the three premises that legitimated it: that communication by itself engenders development; that growth in production and consumption of goods and services constitutes the essence of development and results in a just division of incomes and opportunity; and finally, that the key to increasing productivity is technological innovation, regardless of whom it benefits or hurts. When we take into account the power structure of the societies where this model was applied, the apparently unambiguous concepts of empirical sociology assumed another meaning: the obsession with the "individual" results in a neglect of the weight of social factors in decision making; the notion of "leader" conceals an elite or oligarchy; the "cosmopolitan" disguises the community of interest between rural and urban authorities; and the "reference group" dilutes the reality of relations of power and internal domination of which the peasantry was victim.[33]

This critical perspective is completely borne out by the numerous initiatives of "popular education" which, from the early 1960s, were undertaken in the Third World, notably the literacy experiments and the "consciousness raising" of the Brazilian educator Paulo Freire. To the "banking" conception of rural extension Freire counterposed a "pedagogy of the oppressed,"[34] which began from the concrete situation in which the learner lived and used it as a source from which knowledge progressively emerged, in a two-way encounter between teacher and learner.

This quest for popular participation in the process of development was so strong that it managed to unnerve some theoreticians of modernization before the 1970s were over.

The Panacea of Televised Education

"In a region like India where about three quarters of the adult population are illiterate—almost nine women in ten—and where a considerable number of people live in villages which are not easily accessible, it is not easy to transmit the idea of family planning on a vast scale."[35] This observation made in June 1966 by a U.N. Advisory Mission can be compared to one made by the Brazilian military regime that had taken power two years earlier, and which, by contrast, expressed less worry about family planning than about illiteracy as a burden on the country's modernization project. For the Brazilian centurions, it was clearly unthinkable to take up the educational principles of a Paulo Freire—he had been forced into exile—which would have transformed literacy into a veritable national crusade. In order to be viable, this alternative would have required the political mobilization of the popular classes. Freire's popular pedagogy, after all, incited the learner to reappropriate his or her experience and history; it associated learning with apprenticeship and consciousness raising.

Instead, all hopes were placed in the early experiments in teaching by satellite, whose chosen territories were India and Brazil. Their immediate objective was summed up in the expression "communication for development"; their medium-term target was to transform these two pilot countries into a shop window for the application of space-age technology to the needs of the Third World.

In the protocol of the agreement signed in 1969 between New Delhi and Washington, NASA and the Indian Department of Atomic Energy gave themselves six years to get the experiment under way. The document stated explicitly that, by this means, India hoped to "increase agricultural productivity, support the aims of the family planning policy, and cement national cohesion."[36] In exchange for U.S. aid and scientific and technical support, India formally promised to "evaluate the results of the experiment and put them at the disposition of the entire world." The accord even specified that the evaluation should be made wherever possible in quantitative terms, the aim being to evaluate family planning measured by birth rates, agricultural productivity, and the growth in incomes, comparing villages equipped with television with those that did not have this means of access to modernity.

To the strategists of space technology, India had seemed the ideal country for this type of experiment, since it had practically no television system—in fact, only one channel received by 10,000 sets—and 550,000 villages to be linked. The promotional literature for the project vaunted the enticing prospect that a satellite could complete in only 10 years a task that a conventional system would require 30 years to accomplish, and for an equal rate of annual investment.[37] The experiment was dubbed SITE (Satellite Instructional Television Experiment) and made use of the American satellite ATS-6. It reached about 2,300 villages, belonging to six states, which were equipped with sets for collective viewing. Relayed by the satellite, SITE was inaugurated on August 1, 1976, and lasted one year. From the point of view of its original objective ("stimulating aspirations"), its results were modest: while the experiment favored national integration by subjecting all the receivers to the same programs, and while the participating teachers found in it an opportunity to escape their isolation and improve their teaching practice, the results in terms of improvement of agricultural practices and adoption of family planning turned out to be insignificant, according to numerous evaluations.[38] The display window, once the project expired, lost its sparkle.

Between the agreement and its realization, even if partial, India had the opportunity to air its ambition to transform itself into a space and computing power. A year before SITE came to fruition, a Soviet rocket had launched its first satellite built by Indian engineers. In the late 1970s, Indian television opened up to commercial advertising and its profits shot up. The state proposed a plan under which industrialists could buy half-hour slots of airtime, in exchange for which they would sponsor a serial of 25-minute episodes and have the right to one minute of free advertising. "The first serials produced in these conditions," recalled an Indian journalist, "especially *Rajani* (the portrait of a housewife who is a populist activist) had astounding success: the network was overwhelmed by sponsors' demand."[39] The entertainment function of television gained more and more space, which it would share with video, catapulted into importance by pirating operations organized out of Singapore. In 1990, television would cover more than three-quarters of the territory and almost half the people would watch its programs and videotapes. Only AIR (All India Radio), which had gone from 90 stations in 1985 to 250 five years later (the first channel with a national reach had begun in 1988), continued to assume the educational role of the media.[40] According to some, it is now even in a position to rival television's information function, so strong is the gravitation toward entertainment in television programming.

High Tech Salesmen

The policy of modernization through satellite technology selected as another theater of operations one of the poorest regions of the federal state of Brazil, the Rio Grande do Norte. The central problem here, apparently, was catching up with remedial schooling: more than 40 percent of the school-age population did not attend.[41]

This project, under the direct supervision of the Institute for Space Research (INPE), one of the top institutions for technology policy in Brazil (thanks again to a hook-up with a NASA ATS satellite), was supposed to demonstrate on a reduced scale the effectiveness of a prototype of a "total system" for the use of audiovisual technology in primary education and to provide the example of an organizational model that could be generalized. Accordingly, responsibility for the project fell to specialists in systems engineering management. The specialists were sent to be trained in the space division of General Electric in the United States; they returned to Brazil with the idea that it was possible to transpose this mode of organization to the project in Rio Grande do Norte.

The pilot plan, dubbed SACI-EXERN and launched in 1974, was interrupted in 1977–78 on the official pretext that the cost of a future satellite would be too high. This incident pointed to the contradictions within the Brazilian state between telecommunication strategies, educational strategies, and scientific policy, not to mention the cleavages between civilian and military interests, particularly important in a regime identifying itself with "national security." It meant above all the failure of a technocratic solution to the education question, a failure of what the Brazilian sociologist Laymert Garcia Dos Santos has called the "systemist approach," carried away as it was by the "disorders of rationality."

The duration of the project was nevertheless sufficient to allow one to observe the development philosophy that informed its experts, such as those from Stanford who from 1967 onwards traveled all over the world prospecting future markets, armed with attaché cases containing copies of a report promising miracles: the ASCEND report (Advanced System for Communications and Education in National Development). As Garcia Dos Santos explains, "It is as if Stanford University were the nerve center where the promotion of teleducation satellites radiates forth to all continents, giving this promotion a scientific status and cultural credentials that feasibility studies from the laboratories of firms directly concerned in their manufacture could not claim to furnish."[42] And the same researcher gives the report the coup de grâce: "What strikes the reader of this report is that it has none of the quality of university work, if one understands by this the discussion of the theoretical foundations of a line of reasoning and the effort to analyze and understand a given problem. The ASCEND report is, rather, a viability study whose main ambition is to sell a specific technology. Its authors say so openly: 'Our aim is to show what must be done tomorrow with today's technology so that every nation can incorporate it in its national development plan.' "[43]

This was not, however, the path taken by the Brazilian state. In 1977, relayed by the Intelsat satellite, the private television network Globo, through its educational foundation named after Robert Marinho, the president of the Globo multimedia group, inaugurated its programs of distance teaching, devoting its first daily broadcasting hours to remedial schooling. Globo progressively attracted sponsors as prestigious as the pharmaceutical firm Hoechst and the Banco do Brasil.[44] In 1990, Brazil would be the home of the fourth largest network in the world, possess its own satellite system, and control a project for manufacturing its own space vehicles. But the 1980s ended with an alarming educational record: according to official figures, 42 percent of children left school before completing the primary level; a quarter of the work force consisted of children aged 10 to 14 earning a salary three times less than that of adults. In ten years, adult labor in the Northeast had grown by 12 percent, and that of children by more than 100 percent.

In view of these figures, one may better understand why some survivors of 1960s modernization theory dared to speak, two decades later, of the "revolution of rising frustrations." There is no better way to grasp this than through Jean Baudrillard's text on the "Cockaigne-land fantasies of the ideology of consumption," written in 1968, at a time when empiricist sociology swore by modernization and democracy-by-consumption:

> The excess of aspirations with respect to real possibilities reflects the imbalance, the profound contradiction, of a society in which the "democratic" *ideology* of social progress often comes to compensate for and overdetermine the relative inertia of social mechanisms. In other words, individuals *hope* because they "know" they can hope; they *don't hope too much* because they "know" that this society in fact poses insurmountable barriers to free ascension; they *hope nevertheless a little too much* because they also experience a diffuse ideology of mobility and growth. The level of their aspirations therefore results from a compromise between a realism fueled by facts and an irrealism maintained by the ambient ideology—a compromise which in turn reflects the internal contradiction of the whole society.[45]

Requiem for a Model

In 1976, the pioneer of diffusionism, Everett Rogers, hastened to bury the "dominant paradigm" and proposed to move beyond the ethnocentric vision that had guided him. Referring to Mao Tse-tung, to Paulo Freire's "pedagogy of the oppressed," and to critics from the Third World as well as the First, he admitted the bankruptcy of quantitative conceptions of development and of their communicational logistics, which had only succeeded in further unbalancing an already highly skewed social structure. The new definition of development advocated by the Stanford sociologist no longer had anything to do with the one he had formulated in the early 1960s. Development now became a "widely participatory process of social change in a society, intended to bring about social and material advancement (including greater equality, freedom and other valued qualities) for the majority of the people through their gaining greater control over their environment."[46]

Rogers adhered all the more strongly to this new vision of decentralized development since it seemed to go naturally in the direction of prodigious expansion of light technologies that could only favor an interactive communication model, a "model of interpersonal networks." *Small is beautiful*: thus he moved from the myth of heavy and vertical media apparatuses as the instigators of innovation to the myth of horizontal micromedia whose decentralized architecture favors the active participation of those interested in the adoption of novelty. It was one way of rejoining the debate then in progress on the general crisis of "development."

Development, for whom, by whom, and why? In many places such were the questions raised about a mode of growth that had become an end in itself: "the accumulation of capital with a view to accumulating still more capital," as Immanuel Wallerstein puts it. What began to be framed as a problem was the justification of this socially absurd objective by the long-term social benefits this kind of growth was supposed to procure. The biological notion of growth of the theoreticians of modernization and their "opinion leaders"—growth sustained by exponential rates whose circle of beneficiaries was assumed to continue expanding—was hereafter subjected to serious examination.

"Rely on your own strength" and mobilize local resources to satisfy local needs: such was the line of action proposed by the new philosophy of development to counter the earlier model, conceived as a movement of extraversion, powered by trade and transfers from the outside. This was the strategy of *self-reliance*.[47] At the heart of this quest was the rehabilitation of specific cultures in defining particular paths to development and the inclusion within the notion of "basic needs" of citizens' participation in the production of society. It thus involved reflection on individual and collective solidarity at the local, national, and international levels.

Such a generous idea would give rise to numerous interpretations, inspiring as many state strategies as actions by civil society. Articulated for the first time in 1967 with the Arusha Declaration, the Non-Aligned Movement would adopt this idea three years later, seeing in it a means of correcting the deficiencies and the slowness in application of measures designed to institute a new world economic order. In incorporating it into their doctrine of development, the United Nations consecrated the concept of "endogenous" or "autocentered" development. This concept sparked innovative thought on the "industrial imperative," on the modes of technology transfer, on the cultural models this presupposed, and on the margin of maneuver of a dependent country in its negotiation with the world-system.[48] On the level of international cooperation, this concept would give rise to others, codevelopment and decentralized cooperation, for example—terms reflecting the attempt to define new forms of international relations between North and South. It gave rise as well to scenarios of development-participation that would mobilize, around concrete projects and situations, other actors (the informal sector, associations, local communities) than those enthroned by development-modernization.

Certain specialists of international communication, considering the margin for Third World countries to be too narrow, not to say nonexistent, would propose pushing the idea of autonomy to its extreme of autarky, postulating the necessity for countries of the periphery to "disconnect" or "dissociate" from the world-system.[49] Unfortunately, they did not take into account the lessons of the tragic history of single-party states that chose to remove their peoples from the influence of international networks of communication.

Chapter 8

The International Regulation of Information Flows: Two Colliding Views

The Principle of Noninterference

In which directions do international information and communication circulate? What economic, political, and cultural stakes do they represent in the relations between the major powers, and more generally among the diverse nation-states? How are they regulated? Is it desirable to fix rules? But which rules and how effective can they be? These questions dominated the 1970s, when other actors, other political and cultural reference points than those prescribed by the logic of the Cold War, sought to express themselves in international negotiations.

And yet it was in an umpteenth episode of diplomatic confrontation between the two superpowers that the debate began. In 1969, the experts of the committee on the peaceful use of extra-atmospheric space, created nine years earlier at the behest of the General Assembly of the United Nations, put forth their conclusions: the impact of direct broadcast by satellite (DBS, which suppresses the need for terrestrial relay stations) must be foreseen in the long term; it was therefore necessary to turn to the problem of methods of regulation.

In November 1972, during the 27th General Assembly, the Soviet delegation proposed that an "international convention on principles governing the use by states of artificial earth satellites for direct TV broadcasting" be prepared. It based its proposal on the work of the committee and on the debates that had taken place on its initiative before UNESCO several months earlier. The proposal was adopted by all the delegations but one, that of the United States. The first paragraph of the approved text read as follows: "The activities undertaken in the domain of international direct television by satellite should be compatible with the sovereign

167

rights of states, including the principle of noninterference, and with the right of everyone to research, receive, and spread information and ideas proclaimed in the pertinent charters of the United Nations."[1]

How can the principle of the free circulation of ideas and information, touchstone of the Universal Declaration of the Rights of Man of 1948, be reconciled with the principle of national sovereignty? Since the end of World War II this question had been at the center of the debate on how to reorganize the world. Direct television merely sparked anew the confrontation provoked by the question of jamming radio waves. In 1950, the point of view of those who opposed jamming international broadcasts had carried the day, to the annoyance of the Soviet Union and the Eastern bloc countries, which saw in it a means of fighting against "ideological aggression" and all forms of "Western propaganda." While France, notably during the Algerian War (1954–62), and the United Kingdom, especially during the Suez Crisis (1956), had also resorted to jamming radio programs from Nasser's Egypt, this practice of interference was the norm for years in the Eastern bloc. Alternately flexible and hard-line, according to the fluctuations of the barometer of internal and external tensions, the authorities never ceased trying to prevent their citizens from listening to the "propaganda stations" and, above all, to the official radio station of the U.S. government, the Voice of America, and the two clandestine stations, Radio Liberty and Radio Free Europe, launched by the CIA in 1953 and 1950 respectively, the former with the exclusive mission of bombarding the Soviet Union, and the latter, Eastern Europe, with its message. In 1972, Voice of America broadcast to Iron Curtain countries 185 hours per week; Radio Liberty, 24 hours; and Radio Free Europe, 540 hours.[2]

The scope of Soviet interference with the airwaves was such that U.S. officials claimed that its budget alone was greater than the entire U.S. budget for radio propaganda. In 1960, listening to foreign radio was even sanctioned as an "ideological crime," and for several years Soviet authorities preferred to brake the technical progress made possible by the transistor in order to cut listeners off from international information flows.

The experience of radio jamming during the 1950s and 1960s had nonetheless already shown how porous the system of technical protection against aggressors was. Citizens were listening more and more to foreign radio, as witnessed by the analysis by East German philosopher and economist Rudolph Bahro, who in 1978, twelve years before the fall of the Berlin Wall, predicted:

In the yawning chasm that has been opened, the mass ideological production of the West has poured in. . . . When one day the technology that made satellites possible totally liquidates the Soviet masses' anachronistic isolation from the "image of the current world," the leaders of the apparatus in Moscow will find themselves ruling over a volcano of unsatisfied material needs. This and nothing else is at the origin of the feeling of panic apparent in the project advanced by [then Moscow's foreign minister] Gromyko for a convention on "Principles governing the use by States of artificial earth satellites for direct TV broadcasting," a document which recalls the style of Nicholas I. In the Soviet Union, it is not only a question of warding off "ideological diversion" in the traditional sense; the propaganda machine will soon find itself totally impotent in its struggle against the vision of the "consumer society."[3]

Bahro knew what he was talking about. The German Democratic Republic was in fact one of the first to experience transfrontier television, with the spilling over of the channels from the other Germany. As for East Berlin, its insular position allowed it to receive the channels of British and American troops stationed in Germany. East German television could not disregard the direct competition and it modified its programming to try to retain its viewers. The places that, for topographical reasons, were deprived of this manna of mass culture were dubbed the "valley of the ignorant."

Education Imperiled by Entertainment

The frontal collision between East and West was not only the telescoping of two principles of news organization—that is, on one side, information retention, penury, and secrecy and on the other, saturation and (at least apparent) transparency. This confrontation via audiovisual systems also involved two different conceptions of culture, two ways of envisaging "cultural democratization." On the one hand, there was mass culture as an extension of a concrete political system, that of liberal democracy, as an expression of a project of social co-optation, and as a particular way of producing consensus, more and more dependent on the logic of the market, and as such structured around "entertainment." On the other hand, there was a project in which access to the goods of classical high culture was the distinctive mark of the democratization of leisure, a project indissociable from the idea of the pedagogical mission of broadcasting. As it was summarized in 1973 by a Soviet television official commenting on the importance of music programs among those conceived for a broad public: "While meeting the viewers' requirements for relaxation, pleasure, and entertainment, we must not forget our basic task, which flows from the great mission of television to educate the social consciousness, to raise the viewers' level of cultural education and to form their aesthetic taste. Television must propagate the progressive ideas of our times among the masses and must be the instrument of their spiritual enrichment."[4] Could one imagine a Western television station, guided by the logic of audience maximization, going off the air one day a week to allow viewers to go to the theater or to a concert, or to participate in other cultural activities, as was the case in Hungary?

Reading between the lines, beyond the repressive connotations of the Soviet thesis on the necessity of regulating information flow, there are two opposing ways of conceiving and realizing leisure. Because, on the other side, behind the thesis of "free flow of information," defended tooth and nail by the U.S. delegation in the forums of the United Nations, one could see the same operating principle of mass culture and of media modernity: *entertainment*.[5] Under the guise of the defense of each individual's right to receive all information freely and retransmit it without regard to borders, this principle, as revised by U.S. diplomacy, became the defense of the freedom of unhindered market circulation of cultural products[6]—a market that, moreover, was completely dominated by U.S. corporations. One understands how great was the temptation among certain democrats to condemn both superpowers with equal severity, one for toying with national sovereignty and the other with human rights.

One was decidedly light years removed from Marx's and Engels's utopia, founded on the demolition of the parochial spirit of feudal society. In their *Manifesto of the Communist Party* they had solemnly proclaimed: "In place of the old local and national seclusion and self-sufficiency, we have intercourse in every direction, universal interdependence of nations. And as in material, so also in intellectual production. The intellectual creations of individual nations become common property. National one-sidedness and narrow-mindedness become more and more impossible, and from the numerous national and local literatures, there arises a world literature."[7]

What we should retain from the first diplomatic controversies on the regulation of direct broadcast satellites is basically the following: that new technologies were starting to explode the national regulatory framework; their transnational character was rendering difficult the exercise of sovereignty expressed in constraining laws, while at the same time the oppositions among states rendered difficult the establishment of international conventions. At best, states were prepared to accept "declarations of principles" of more moral than juridical value. In the absence of a solid ground of understanding, the only possible form of political agreement was the "code of good conduct."

This assessment would be reconfirmed in November 1974, when the U.N. General Assembly was invited to discuss the regulation of observational satellites. These satellites had begun their career two years before, with the launching by NASA of the first ERTS (Earth Resources Technology Satellites). The civilian use of satellite images was already of intense interest to industries and researchers as different as oil companies, mineral prospectors, grain growers, agronomists, oceanographers, geographers, and territorial planners, and everything pointed to an inexhaustible field of applications for the techniques of spatial detection. In the course of the first debates on satellite-generated information, France—which would not become a force in this market until the following decade with its Spot satellite—combined with the Soviet Union to propose that information collected on natural resources should not be delivered to a third party without the consent of the observed country. Benefiting from a de facto monopoly on this technology, the United States invoked once more the principle of free flow of information and that of "first come, first served," and asserted its postion in practice: the acquisition of data without prior authorization and its free and non-discriminatory dissemination. When the commercialization of this type of information began, the notion of a "code of good conduct" would be trotted out again.[8]

In 1979, during the World Administrative Radio Conference (WARC), held under the auspices of the International Telecommunications Union (ITU), the U.S. delegation tried without success to oppose the rediotribution of the world radio frequency spectrum, of which certain wavebands were practically monopolized by the industrialized countries to the detriment of the Third World, by invoking yet again the principle of "first come, first served."[9] The conference also marked a turning point. For the first time in the history of telecommunications, the formerly colonized countries of Africa and Asia made themselves heard: 142 nations participated. At the two preceding conferences, one held in Atlantic City in 1947 and the other in Geneva in 1959, at which the principle of priority or the "right of previous use" had been ratified, only 76 and 96 delegations had attended, respectively.

The debate on satellite images was echoed in another controversy that took place between 1973 and 1982 over the notions of "freedom of the sea" and the "common heritage of humanity" with a view to revising the law of the sea with respect to the sovereignty of maritime countries. Legal scholar Monique Chemillier-Gendreau has commented that "the law governing a social field, whatever it is, may be really explained only by what is at stake." And the stake of "geoinformation" is great on a planet governed by the geoeconomy. Chemillier-Gendreau recalls the maxim of

Lacordaire: "Between the strong and the weak, it is liberty that oppresses and the law that sets free."[10]

New Sites of Critical Thought

The debate on the regulation of international flows became more complicated with the emergence of the countries of the South. In the 1950s and 1960s, two determinisms, that of technology and that of modernization, had denied Third World countries the status of major actors in the theoretical schemas of development and growth. Future receptacles of a progress introduced from outside, these societies, labeled "traditional," were reduced to waiting for the revelation of the *dei ex machina* charged with spreading the good cosmopolitan word. There was a mirror-and-screen effect: development-modernization theory incited societies on the one hand to see the image of their future in the ideal model embodied by modern societies of the urban and industrial North, and on the other to consider their own cultural heritage as a handicap on the road to social and economic evolution.

The first signs of a solid critique of this worldview came from Latin America: the region that, in the modernizers' scale of things, had already crawled up the most rungs toward the Promethean objective of "development." The debate on direct television had seen a collision between the nation that was the symbol of informational opulence (the United States) and the one that had made it a rare good (the U.S.S.R.). The other debate over the imbalance in flows of communication began by confronting a region whose experience of daily commercial mass communication was as old as that of radio and television itself, with the power which at the time concentrated, according to Zbigniew Brzezinski, more than 65 percent of the world's total flow. Latin America alone accounted for two-thirds of the Third World's media resources. In addition, its dominant model for the organization of audiovisual media was the closest to the archetype inaugurated by the United States. While Western Europe was still entirely under the regime of public service and public monopoly, the overwhelming majority of Latin American countries, had already lived for several years under the sign of advertising overstatement, and the logics of competition and internationalization of programming. And yet they considered themselves "underinformed."

To the pioneering critical studies of the theories and strategies of modernization were thus added analyses of the dependence of national media on foreign sources of news and programs. The context of the long history of conflictual relations with the United States since the end of the nineteenth century helped to explain this situation. One of the first works on news flows was the book by a journalist and professor from Venezuela, Eleazar Díaz Rangel, longtime president of his country's association of journalists. Published in 1966, it was entitled *Pueblos sub-informados* (Underinformed peoples) and its point of departure was the news put out by the large news agencies, particularly UPI and AP, during the landing of the Marines in the Dominican Republic the previous year.[11] It was also in Venezuela, a country characterized by an extremely commercial television and an already well-developed advertising industry, that the break began with visions arising out of North American empirical sociology, then hegemonic throughout Latin America. The principal works of that sociology had been translated into Spanish thanks to the support of official U.S. agencies, and were distributed free of charge to universities by the USIA. It was under the direction of researchers such as Antonio Pasquali, strongly influenced by the Frankfurt School, that the first analyses of television as a cultural industry were carried out, starting in 1963.[12]

Two other poles of critical research sprang up, before the decade was over, in Argentina and in Chile. In these countries began research that departed from the beaten paths of both North American empiricism and an orthodox Marxism incapable of treating the media other than as dispenser of propaganda or pedagogical vehicle. Marked from the start by the French school of structural semiology, and in the case of Chile also by the Frankfurt School, this new type of approach to the media was forced into breaking with its theoreticist tendency by rapid political developments in these countries.[13]

In the context of a determined opposition within a pluralist democracy, critical researchers in Chile posed new questions on the nature of mass culture—which had in fact become daily culture, having entered into the reflexes and patterns of life. They understood why Roland Barthes refrained from speaking of "propaganda" and preferred to call it "mythology" instead. During the three years of the Chilean Popular Front experience (1970–73), the clash between mass culture and the project of social change was constant. The new questions raised concerned the difficulty of redefining the emotional relation between the media and their users; the difficulty of imagining forms of participation other than the "sensorial"; the difficulty, thus, of finding forms of democratic control over the media; the difficulty of linking the media to a project of social development that broke with the mirages of the "revolution of rising expectations"; the difficulty of mastering flows of a communication more and more tied into a transnational structure of information production (as was shown by the conflictual relation between the popular government and the major news agencies, such as UPI, which too often only reproduced internationally the news appearing in the dailies and magazines of an openly seditious oppositional press). Finally, there was the inability of traditional political schemas to take account of the role played by the media in a society politically and socially divided in the extreme. These were the issues that the coup d'état against Salvador Allende in September 1973 left unresolved.[14]

History French-Style

The conviction that it is less and less possible to treat the media and communication without tackling the logics of internationalization would during the 1970s inspire a number of research centers, not just in the Third World but also in the First.

Thus, in the United States, in 1969, Thomas Guback, professor at the University of Illinois, published a study on the international film industry in which he analyzed the balance of forces between Europe and the United States in the area of film since 1945. In his conclusion, he warned European governments:

> Twenty or even fifteen years ago, Europeans undoubtedly did not realize the consequences of their open arms policy. They saw foreign involvement as an aid—but were slow to recognize it as a danger. . . . If economic independence and cultural integrity are to prevail, then European industries and governments must respond to the two thrusts of the American industry—production financing and the international distribution system. . . . Independence does not necessarily mean "better" films in an artistic or financial sense, any more than internationalization means "better" films. But autonomy can increase the chances for diversity and different points of view.[15]

The same year, Herbert Schiller of the University of California published his first book, *Mass Communication and American Empire*, in which he analyzed the industrial complex of communication of his country and established a direct link between its spectacular takeoff and the ascendancy of military interests.[16]

Other breeding grounds for international communications research were Great Britain and the Scandinavian countries. Two Finnish researchers, Karl Nordenstreng and Tapio Varis, carried out the first study commissioned by UNESCO on the import and export of television programs.[17] From Norway, in the framework of peace studies, came one of the first researches of the structure of international news,[18] a theme that also mobilized researchers at the Centre for Mass Communications Research at Leicester University, under the direction of James Halloran.[19]

Among the first studies published by the researchers at this British center, aside from the critiques of the theories of modernization-development already cited, was a well-researched investigation of the way in which the media handled a peaceful demonstration against the war in Vietnam that took place in London in October 1968, and a study of the press coverage of international affairs.[20] The Scandinavians and the British were thus among the first (and rare) teams to tackle the relation between war (including psychological warfare) and the media.[21] The 1960s would see a proliferation in the English-speaking countries of research on international communications, stimulated by the debates taking place in the major international organizations.[22]

The rise of new centers of criticism is contemporaneous with the new approaches proposed by development economics, and specifically with the idea that it is impossible to understand the history of modern capitalism outside the context of the world system to which it has given rise. For this conception of world integration, the process of underdevelopment is only explainable through the history of relations of structural dependency linking the "central nations" with the "peripheral nations." Launched by Paul Baran in 1957 in his book *The Political Economy of Growth*, this hypothesis would see numerous variants differentiating themselves according to the degree of autonomy attributed to the dependent country by the world system in the search for a model of national development.[23] But all agreed on the need to criticize the evolutionist vision of the exponents of development-modernization.

Above and beyond the strategic and tactical differences that divided the diverse representatives of this new theory of dependency, and which would appear more clearly two decades later when the reform/revolution alternative was abandoned as irrelevant, what these writers proposed was an alternative vision of the formation of the world market. They restored to capitalism its dimension as a historical system, a global system of production and exchange whose market networks have woven ever more tightly together the economic, political, cultural, and scientific spheres, as well as the local, national, and transnational levels. It must be immediately pointed out, however, that many economists and historians of dependency, carried away by their analyses of the supranational dimension of the growth dynamic of this system, minimize both the extra-economic dimensions and the subnational ones.

The American historian Immanuel Wallerstein, in the lineage of the idea of world economy offered by the French historian Fernand Braudel, suggests what the concept of world system has contributed to thinking on the genesis of communication networks:

Were [commodity chains] all plotted on maps, we would notice that they have been centripetal in form. Their points of origin have been manifold, but their points of destination have tended to converge in a few areas, that is to say, they have tended to move from the peripheries of the capitalist world-economy to the centers or cores. . . . The real question is why this has been so. To talk of commodity chains means to talk of an extended social division of labor which, in the course of capitalism's historical development, has become more and more functionally and geographically extensive, and simultaneously more and more hierarchical. This hierarchization of space in the structure of productive processes has led to an ever greater polarization between the core and peripheral zones of the world-economy, not only in terms of distributive criteria (real income levels, quality of life) but even more importantly in the loci of the accumulation of capital.[24]

The vigor of this fundamental movement in international communication studies in English-speaking and Latin American countries contrasts with the hesitant attitude of French research. Apart from a few isolated studies—one of which was published by Hervé Bourges (who later became the director of French public service television)—French research has been largely absent from the debate on the stakes of the internationalization of communication throughout the decade.[25] This gap merely confirmed in this field what others had noticed in their own disciplines—history, for example. As Michel Vovelle writes:

There is widespread agreement on the excessively narrow and Francocentric character of French history, the abandonment of European history, especially that of the northern and eastern countries, because Mediterranean studies are in better shape, with those on Spain and especially Italy never having stopped, in the wake of Braudel, capturing the attention of French researchers. Above all, we see the poverty of non-European history, whether it be of the United States or the Third World, in spite of brilliant exceptions.[26]

The deficiencies noted in the research on international communications in France were all the more serious in their consequences in that, already in the 1970s, understanding the significance of the state's preoccupation with new information technologies became difficult, if one ignored the new industrial and cultural challenges launched on a world-space scale. This lack of dynamism corresponded to the closed-mindedness of French cultural diplomacy as well as of French private cultural industries. In 1979 Jacques Rigaud, a high French official who was soon to become head of the only multinational television channel at the time (RTL), wrote in a report to the minister of foreign affairs:

> The interdependence of cultures is no longer a theme for philo-sophical thought but a lived reality. Dominant models, transmitted via ideological or economic imperialisms, or simply via the stan-dardization of customs, create value references planetary in scope. There follows from this a tendency visible throughout the world, both to exalt the cultural identity of nations, local communities, minorities of all kinds, and to recognize an emerging universal civi-lization. . . . France is moving quickly away from her tradition of cultural internationalism. . . . We are falling back on the national space while believing we still radiate throughout the world.[27]

Media Imperialism: A Reductive Concept

Ideological imperialism, economic imperialism, cultural imperialism: since the end of the 1960s, these terms, used both by a Jacques Rigaud, alarmed about the loss of French cultural influence in the era of informa-tion technologies, and by a Zbigniew Brzezinski,[28] who believed them outmoded, have run through studies on the role of communications in the relations among nations.

Some scholars have even proposed grouping under the heading of "media imperialism" the different currents of critical research on interna-tional communication. Among them was British scholar J. Oliver Boyd-Barrett, who defines it as "the process whereby the ownership, structure, distribution or content of the media in any country are singly or together subject to substantial external pressures from the media interests of any other country or countries, without proportionate reciprocation of influ-ence by the country so affected."[29] This definition has been criticized for imputing an intentional character to the process, and thus for not grasp-ing the normality of a mechanism that functions by itself and without anyone lending a hand, except in exceptional periods of crisis and open confrontation when propaganda takes over from the metabolism of a system. It has also been objected that Boyd-Barrett's definition is much too narrow to account for the multiplicity of forms taken by power rela-tions among the various cultures. Herbert Schiller, among many others, prefers the notion of "cultural imperialism," which he characterizes as follows: "The concept of cultural imperialism today [1976] best de-scribes the sum of the processes by which a society is brought into the modern world system and how its dominating stratum is attracted, pres-sured, forced, and sometimes bribed into shaping social institutions to correspond to, or even promote, the values and structures of the dominating center of the system."[30]

In any event, both terms were handicapped by the negative connotations of the Leninist theory of imperialism, even though the new sociopolitical approach represented without a doubt a break with that tradition, which conceived modern imperialism as a stage in the development of capitalism, reducing explanations solely to the economic factor. Empiricist sociology, only too glad to rely on its habit of simplistic binary reasoning and little inclined by nature to epistemological questioning, caricatured this innovative current but did not bother to examine closely why and how, in various places in the world, a common need had emerged, specific to each situation, to find other ways of seeing the international relations between cultures occupying very different positions with respect to the axes of world power.[31]

During the 1970s, nevertheless, direct confrontations within international organizations over the unequal exchange of flows resulted in the hardening of positions on both sides, pushing into the background any discussion of the complexity of international relations. This complexity had already been noted by Antonio Gramsci in the late 1920s in his struggle against economic reductionism. As he wrote in the *Prison Notebooks*,

> It is necessary to take into account the fact that international relations intertwine with these internal relations of nation-states, creating new, unique and historically concrete combinations. A particular ideology, for instance, born in a highly developed country, is disseminated in less developed countries, impinging on the local interplay of combinations. This relation between international forces and national forces is further complicated by the existence within every state of several structurally diverse territorial sectors.

He took his illustrations of international actors from the cultural and ideological circuits with which he was most familiar:

> Religion, for example, has always been a source of such national and international ideological-political combinations, and so too have the other international organizations—Freemasonry, Rotarianism [which appeared to Gramsci to be one of the important networks for the transmission of Americanism], career diplomacy. These propose political solutions of diverse historical origin, and assist their victory in particular countries—functioning as *international political parties* which operate within each nation with the full concentration of the international forces. A religion, Freemasonry, Rotarianism, etc. can be subsumed into the social category

of "intellectuals," whose function, on an international scale, is to mediate the extremes, of "socializing" the technical discoveries that provide the impetus for all activities of leadership, of devising compromises between, and ways out of, extreme solutions.[32]

These observations are intelligible only if one recalls that, for Gramsci, the term "party" has a much broader meaning than the one attributed to it by political science or by common usage; it overlaps with the meaning of "organizer" or "organic intellectual" and is inseparable from the concept of hegemony. Gramsci's work, then, already invited an analysis of networks for production of consensus and systems of alliance on an international scale.

This necessity of taking into account the mediations and the mediators in the meeting between individual cultures and world-space would be stifled by the ideological polarizations that led to seeing "blocs" where there was in fact diversity, smoothness where there was roughness, simple equations where there was cultural complexity, and one-way traffic of meaning where there was circulation.

In 1983, in search of an alternative perspective to account for the links between local, national, and transnational dimensions in communication processes, Michèle Mattelart, Xavier Delcourt, and I wrote in the Introduction to our *International Image Markets*:

> To respond to this proliferation of questions, the notion of cultural imperialism, and its corollary, "cultural dependence," is clearly no longer adequate. Historically, these two notions were an essential step in creating an awareness of cultural domination. Thanks to this consciousness, a political and scientific groundswell was progressively built up, intimately linking subjective impressions acquired in day-to-day struggle with attempts to formalize a theoretical base. Without this lived experience, it is impossible to understand the hesitations, uncertainties and also the conceptual certainties of diverse geographical and social areas. One day, we shall have to examine more closely not only the genesis of communication systems but also the history of the manufacture of the concepts which made them into a privileged area of research. Only this recording of history enables us to seize at once the continuities and the ruptures which have given rise to new approaches, new tools, linking up with real social movements.[33]

The "New Order": Dialogue of the Deaf

In 1969 UNESCO, then presided over by the Frenchman Jean Maheu, convened a meeting of experts in Montreal on the request of its members.

On the agenda was an overview of research and an outline of probable tendencies with a view to adopting a strategy of support for research in the years to come. The working document distributed to the participants was drafted by the Englishman James Halloran. This meeting called attention to the inequality of the world division of research, centered as it was in large part around certain themes and the situations of industrialized countries. This inequality was a direct reflection of other imbalances in the areas of economics and information. One passage of the meeting's conclusions reads as follows:

> At the present time, communication takes place in one direction. . . . The image given of developing countries is often false, deformed, and, what is more serious, this image is the one presented in these countries themselves. The participants in the Montreal meeting believe that the exchange of information and of other cultural products, particularly in developing countries, is in danger of modifying or displacing cultural values and of causing problems for the mutual understanding among nations.

The idea that this inequality in information flow must be remedied began to gain currency.

In 1973, the fourth conference of heads of state of the nonaligned countries, held in Algiers, proposed the institution of a "new international economic order." This proposal was ratified by the U.N. General Assembly in May 1974 and a calendar of measures was drawn up. The U.N. assigned itself the goal of countering the deterioration in the already unfavorable terms of exchange for developing countries who were producers of primary products; negotiating unlimited access to the market of developed countries; reinforcing financial flows by revising the protocols of access to credit from the International Monetary Fund (IMF, the most important credit institution in the world monetary system); favoring Third World participation in the management of the IMF; establishing a code for technological transfers; and more generally, promulgating a code of conduct for multinational corporations. Finally, the U.N. committed itself to taking measures to facilitate the redeployment of industry toward the countries of the Third World.[34]

The Algiers summit also recommended the reorganization of communication systems in the nonaligned countries themselves. In 1975, on the initiative of the Yugoslavian agency Tanjug, the first pool bringing together news agencies from the nonaligned countries was born, with ten agencies belonging. At Tunis in 1976 the necessity of "decolonizing information" was proclaimed at a symposium organized by the Non-Aligned Movement. The same year, the fifth such conference was held in

Colombo, Sri Lanka, and its contribution was to launch the idea of a new world information order, later known as NWICO, the New World Information and Communication Order. This demand, presented as an indispensable complement to the inauguration of a "new world economic order," would be incorporated into the programs of UNESCO and the U.N. General Assembly in 1978.[35]

In the course of numerous meetings, the main target was defined: the four or five major press agencies, "world agencies," which handled about 80 percent of the information destined for the public—that is, the two principal European agencies, Agence France Presse (AFP) and the British agency Reuters, and the two American agencies, Associated Press (AP) and United Press International (UPI). The accusations leveled at them from the forums of the Non-Aligned Movement unleashed a spiral of verbal violence in the major Western newspapers and in a number of commercial press organs of the Third World whose editors were in solidarity, against their own governments, with the First World. In any case, the debate on the new information order greatly mobilized the U.S. communications industry, which not only interpreted it as a dangerous precedent for the survival of freedom of the press but saw in it a real threat to the principle of the free flow of information, the basis of the future information society.[36] For while the large news agencies were presented as the bogeyman, the stakes of these debates were obviously elsewhere. But these stakes were never publicly made clear by institutions like UNESCO, the major arena of a confrontation that brought together two sorts of polarities: North versus South and East versus West, the latter overdetermining the former. The East ably succeeded in fusing its position on the responsibility and thus the intervention of the state in defense of national sovereignty with that of countries of the Third World fighting for their cultural self-determination.[37]

The report of the international commission for the study of communications problems, presided over by the Irishman Sean MacBride, founder of Amnesty International and winner of the Nobel and Lenin Peace prizes, did not succeed in softening entrenched positions on either side.[38] This commission, created in 1977 by Amadou Mahtar M'Bow, the Senegalese successor to Jean Maheu as director general of UNESCO, offered all the guarantees of pluralism. It included among its 16 members personalities as different as Hubert Beuve-Méry, founder of the newspaper *Le Monde*, the Colombian novelist Gabriel García Márquez, the director of the Soviet press agency Tass, and the Tunisian Mustapha Masmoudi, spokesman for the nonaligned countries. The analyses contained in the MacBride Commission report, whose final version was published in 1980, fell well short of the large quantity of academic research and official reports already in circulation at the time. Not only were its proposals for realizing a new order mere generalities, but the diagnosis itself was hardly prospective.[39] Scarcely any place was given to many hypotheses then in circulation, formulated from very different ideological and philosophical positions, on the implications of the international rearrangement of communications and information systems—from the one put forward by Brzezinski to those of Nora and Minc, as well as the new critical research on the political economy of the media.

In any event, the MacBride report made everyone unhappy, even if one cannot deny its distinction of having been the first official document published by a representative international body in which one finds the question of the inequality of information flows posed in black and white. Scarcely 15 years earlier, UNESCO experts did not even touch on this question, absorbed as they were by the calculation of indices and models of modernization. Unhappy with the turn taken by the debate, the Reagan administration, soon followed by Margaret Thatcher's Great Britain, walked out of UNESCO in 1985. Two years earlier, the U.S. Senate had established its position in a report (see chapter 6 above). Condemning the "politicization" of international institutions such as UNESCO and the International Union of Telecommunications (IUT), the Senate report recommended that Washington remedy the situation by "assuring efficient nonpolitical international organizations for the development, management, expansion and nondiscriminatory access to international telecommunications facilities and networks."[40] Thus the Senate reminded the U.S. government that it could not allow itself to dissociate problems of "cultural information" from those regarding the expansion of telecommunications networks.

The Lack of Moral Credibility

The debate on the new order ran into an impasse that cannot be explained solely by the intransigence of the neoliberals. Another decisive factor was the lack of moral credibility of certain protagonists of the debate.

In 1980, the Venezuelan researcher Oswaldo Capriles, one of the first to recognize the need to remedy the inequality in flows, noted

> the excessive predominance of Third Worldism as a justification at any cost of a struggle that does not often appear to distinguish between democratic and progressive states on the one hand, and totalitarian or reactionary states on the other. Thus countries with feudal political regimes, enemies of human rights, appear to be on the side of countries that are making a real effort to progress in the

economic, political and cultural liberation of their peoples. . . .
Many countries of Latin America and the Third World have taken
advantage of the new world information order as a *fuite en avant*
to abandon the demanding and dangerous field of national policies, arguing that the international field should take priority. The
ardent defense of a new technological order is, quite often, a convenient disguise serving to maintain the internal situation un-
changed.[41]

Indeed, examples were not lacking of governments that, while taking
the lead in demanding a new communication order and creating agency
pools in the name of cultural identity, did not shrink from muzzling the
press *in domo*, imprisoning journalists, and banning from the large or
small screen their filmmakers, who were obliged to go into exile. Nor was
the memory so distant of those local elites who, ashamed of the musical
expression of the popular classes, ridiculed it right up until the moment
when this music, consecrated on the international market, returned in triumph to its native country. Such was the case with reggae. "The middle
and upper classes who controlled the cultural destiny of the country," recalls Caribbean author Sebastian Clarke in his history of rasta music,
"defended their inferiority complex by putting out on the airwaves the
musical stereotypes of their American heroes, and by announcing their
contempt for the 'noise' made by homegrown artists. Since Jamaicans in
general, and the poor especially, did not possess an example of 'culture'
or of 'history,' the dominant groups could not admit that those people
had something significant to say to them.[42]

The debates on the new order ran up against a twofold obstinacy: the
refusal by certain countries of the South to broach the problem of old political censorship exercised by the state in their domestic space, and the
matching refusal by the major industrial countries to raise the issue of the
new economic censorship stimulated by the concentration in the communications industries. Both parties were careful not to raise the central
question for the establishment of more democratic rules for freedom of
expression: How to produce information, in the North as well as the
South, from perspectives other than those of power?

The new order could serve as an alibi, as economists of development
also learned from the evolution of the new world economic order. "The
internationalization of problems of 'development,' " explained Lebanese
author Georges Corm in 1980, "provided a very good alibi to Third
World governments, as a way to invoke to their frustrated populations,
the impossibility of reforming the international economic order, which
aborted domestic 'efforts' at development. For their part, the govern-

ments of industrialized countries found facile themes for their public opinions in the rise in oil prices, immigrant workers and competition from the newly industrialized countries."[43]

But the fact that governments resorted to the argument of the new order to abdicate responsibility for their own unprincipled behavior did not invalidate the existence of the enormous imbalances they denounced, or the pressure to resolve them.

Toward Free Trade

Who could have foreseen, during the earliest discussions on direct broadcast satellites, that 18 years later the Soviet press agency Novosty would ally itself with a major advertising network in the United States to establish a common subsidiary in Moscow? In an irony of history, the principle of self-determination and national sovereignty with which the Kremlin had tried to resist pressure from the United States delegation, now served the various republics in dismembering the Soviet Empire and unleashing ethnic nationalisms. The polemics of that time would appear decidedly remote when the Berlin Wall fell, along with other symbols of the informational closure of "real socialism," and when multimedia groups and advertising conglomerates from the industrial countries rushed into this new frontier opened to the world market.

In the 1980s, regulation and public intervention were no longer on the agenda either in most of the countries that during the previous decade had struggled for a new information order. The state withdrew and Washington's ominous prophecies at UNESCO about what would happen if the state regained control over communication disappeared over the horizon, to the benefit of the private sector. However, the problems raised by the debates of the 1970s remained.

For while it is true that new audiovisual powers of international scope have arisen in countries such as Brazil (with Globo) and Mexico (with Televisa), television for the great majority still relies on the images of others. And while schooling is in decline and illiteracy reconquers ground in a great number of debt-ridden nations that have made huge cuts in their education budgets, the uses foreseen for the communications technologies across the world lean largely toward the "logic of entertainment."[44] Meanwhile, the monetarist strategies for economic development begin to provoke a redistribution of power between the public and private sectors in the realm of telecommunications. The more and more frequent sale of telephone companies in certain regions of the Third World generated substantial revenues that partly reduced large deficits.

The issue of the regulation of international networks has not, however, disappeared from the major international forums. It has merely moved toward more technical bodies, such as GATT (General Agreement on Tariffs and Trade), which since 1947 watches over free trade. Its role has grown considerably in the area of communications since the opening of the Uruguay Round in September 1986, when negotiations on the international trade in "services" were first put on the agenda. Banking, insurance, tourism, and transport are classed in a category along with communication in this accounting of "invisible flows": included are advertising, marketing, telecommunications, the wide range of the products of the cultural industries, not to mention the complex network of data banks and data bases with their multiple uses.[45] The desire to institute free trade in services has a twofold strategic significance. The abolition of obstacles to trade (nontariff barriers, diverse state subsidies and supports) is one of the essential phases in the construction of the world-space. The liberalization of flows is at the very base of the new mode of organization of the corporate network.[46]

These technical discussions are dominated by commercial confrontations that often cut across North-South lines. Moreover, the South is no longer what it was when, in 1952, the demographer Alfred Sauvy and the anthropologist Georges Balandier created the unitary notion of "Third World." The gap has widened between the "newly industrializing countries" and the mass of others. Certain analysts, such as the Indian economist Chakravarthi Raghavan, are rather pessimistic as to the impact of this deregulation of invisible flows on North/South relations and go so far as to interpret it as a "veritable recolonization of the Third World."[47] The creation of vast regional free trade zones, which associate Third World and developed countries, results in a continual reshuffling of the world order, organized around the market.

This rise of market logic and competition across borders has, in turn, nearly ten years after the cry of alarm from the countries of the South, belatedly obliged the European Community and the Council of Europe to debate measures to reestablish balance in the trade of TV programs. Should a quota of European programs be imposed or not? After five years of deliberations (1984–89), the "zero option" defended by Thatcherite England against France (which had recommended instituting a quota to defend a hypothetical "European cultural identity"—and its national industry) finally won the day. With deregulation of national audiovisual systems, the old continent became the geographical unit with the largest deficit in the world in terms of imports over exports, but still more in absolute terms, since it is the world's foremost importer of television programs and the best client of the U.S. program industry.[48]

The influence acquired progressively by the market as a space of reorganization and regulation of world space was underestimated by the futurologists in the 1960s and 1970s. As state servants, they were busy developing anticrisis strategies based on the voluntaristic intervention of the public sector.

PART III

Culture

Chapter 9

The State in Its Ordinary Dimension

Cultured Distrust

"Power escapes from governments and national states in three directions: to local groups who want to act more at their own discretion, to private businesses who want to act more quickly and flexibly than public authorities, and to international organizations, who must somehow try to manage the new technologies that transcend national jurisdictions. In short, governmental institutions are the vestiges of the era for which they were conceived—an era of blind growth during which the multiple and diverse forms of growth were independent from one another."[1] So declared the American political scientist H. Cleveland in the *NATO Review* in late 1978.

What role fell to public power in the regulation of communication in the space traced out by the nation-state? This question would recur throughout the 1970s, in countries with a strong centralizing tradition as well as in those where regulation by the market had already much reduced the jurisdiction of public authorities in this area.

In an apparently paradoxical manner, while the entry of the nation-state into the world-space was accelerating, along with the posing of more and more political and economic problems in these terms, critical theory felt the need to reconsider the specificity of the state and the apparatuses of national as well as local communication. The conceptual tools that had prevailed up to then in accounting for the functioning of national communication systems proved to be inadequate. A twofold explanation of this is necessary.

The first reason is to be found in the way in which the critics of mass culture—and along with them, the majority of oppositional intellectuals everywhere up to the 1970s—had until then approached the process of the industrialization of culture. The first critical theory, conceived in the late 1940s, is the work of the Frankfurt School, and more particularly of Theodor Adorno and Max Horkheimer, two German philosophers exiled to the United States to escape Nazism. From what was in fact the first theoretical confrontation between European Enlightenment culture and mass culture produced "for the millions" resulted the concept of the "culture industry." In a seminal essay, "The Culture Industry: Enlightenment as Mass Deception," the two authors wrote:

> Interested parties explain the culture industry in technological terms. It is alleged that because millions participate in it, certain reproduction processes are necessary that inevitably require identical needs in innumerable places to be satisfied with identical goods. . . . Furthermore, it is claimed that standards were based in the first place on consumers' needs, and for that reason were accepted with so little resistance. The result is the circle of manipulation and retroactive needs in which the unity of the system grows ever stronger. No mention is made of the fact that the basis on which technology acquires power over society is the power of those whose economic hold over society is greatest. . . . A technological rationale is the rationale of domination itself. It has made the technology of the culture industry no more than the achievement of standardization and mass production, sacrificing whatever involved a distinction between the logic of the work and that of the social system. This is the result not of a law of movement in technology as such, but of its function in today's economy.[2]

In Adorno's and Horkheimer's eyes, the culture industry, as a place where serialization, standardization, and the division of labor are carried out, exemplifies the bankruptcy of culture and its fall into the status of a commodity. The transformation of the cultural act into a value abolishes its critical power. The reign of pseudo-individuality, which began with the existence of the bourgeoisie itself, is deployed arrogantly in mass culture. "What is individual is no more than the generality's power to stamp the accidental detail so firmly that it is accepted as such. The defiant reserve or elegant appearance of the individual on show is mass-produced like Yale locks, whose only difference can be measured in fractions of millimeters."[3]

The link they established between technology, culture, power, and economics is not analyzed as such; it is there only to clarify what they consider to be the degradation of the philosophical and existential role of culture as authentic experience. Therefore we must not expect to find in the work of these pioneers analyses of the way in which, in each concrete instance, the cultural industry—used in the singular, because viewed as a total system of production of culture as commodity—is incorporated into the action of institutions, or how it is positioned in relation to the state and organized civil society. And still less should we expect analyses of the way in which each of its components (cinema, music, press, radio, and so on) passes through, in its specificity, this process of industrialization. The notion of culture industry serves therefore as a foil to a certain sacralization of art and of high culture more than it serves to elucidate the industrialization of culture; still less is its internationalization examined. Hence it presents a crudely totalizing thesis. The presence of an industrial mode of production leads the two philosophers to fear that literature, painting, and jazz will meet the same fate as comic strips, radio, and the cinema. And yet, ten years before this essay on the culture industry appeared, Walter Benjamin, another representative of the Frankfurt School, had indicated how the very principle of reproduction renders outdated the old conception of art, which he called "auratic." He showed clearly how an art like the cinema can exist only at the stage of reproduction and not that of the unique production.[4]

What Adorno and Horkheimer seem finally to have refused is above all this reproducibility of a cultural artifact by technical means. One cannot help seeing in their thought a certain nostalgia for a cultural experience free from any relation to technique. As I wrote with Jean-Marie Piemme in 1979: "One finds in this thought a Jansenist conception of writing, which, confident in itself, always suspects other means of communication (notably the image) of being a bearer of Evil. Writing, the repository of originality, is by the same token seen as the guarantor of authenticity and of the rationality of communication. On the other hand, the image intimately linked to reproducibility will always be pregnant with an unwanted irrationality." This "literate distrust" led to the conclusion that "curiously, this kind of value judgment may be found as well at the foundation of other, diametrically opposed, approaches. One finds it in Ortega y Gasset as well as in Adorno: in their cases the weight of cultural heritage overdetermines the system of political and philosophical values."[5]

From this vigorous declaration of faith in the value of high culture came two misunderstandings. With time, it would be severely reproached for bedding down with elitism and precluding the right questions on democratization and cultural democracy from being raised in the age of mass communications. Another objection reinforced the first: Can one legitimately infer from standardization and serialization of the cultural product a "mass production of the individual"? This is an old question that already divided empiricist sociology and that, in the course of time, would also divide different critical outlooks.

The Leviathan

The other factor that explains the inadequacy of time-honored analytic schemes to the new circumstances of the 1970s lies in the concept of the state prevalent at the time. Critical theory postulated an "essence of the state," that is, a stylized and totalizing model of the state, a universal model applicable in all times and places, a rigorously functional Platonic entity.

This abstract and theoreticist conception, adapted to the analysis of the media defined as part of what Louis Althusser dubbed "ideological state apparatuses," eliminated from the field of observation the diverse forms taken by media institutions in each concrete situation. To begin with, a regime of private property is necessarily different from one governed by the principle of public service. The idea of structural invariance is at work here again, not only in the approach to state agencies but also in the analysis of discourses, inspired by the semiology of the era. Certain generic expressions were in vogue: massification, standardization, the dominant discourse, the consumer society, and so forth.

What was neglected in these theories of the cultural industry as a total system and of the state as a metaphysical entity was the historical dimension, that is, the articulation of the media with the whole ensemble of contradictions and structures in which they are implicated. What is the organic link tying a medium to the historical era and geographical space in which it functions? Is it the relation among the media themselves, both within a country and on the international level? Or rather among the economic and political determinations that, at a given moment, leave their mark on the social functions and uses of information and communication technologies?

As long as these questions were left unasked, internationalization was a dimension with no place in these schemas. To the forgetting of time was compounded the evacuation of space. But without these dimensions, there was scarcely any way of conceiving the successive ruptures the state had experienced in its forms, structures, and functions, from the city-state, the empire-state, and the feudal military state to the nation-state. The latter could not avoid coming to terms with worldwide space. We should also take into account the fact that the essentialist conception of the state, like the vision of culture as high culture, had been tainted at birth with another vice: logocentrism, that is, a belief in the universal value of rational models, truth, and normativity, at the foundation of the Western logos. What is most incredible is how these theoretical schemas—which invited political systems as different as Jacobin France, the U.S. federal state, Brazil under military dictatorship, or Mexico under its fifty-year-old single-party regime, all to identify with a universal notion of politics—created illusions for decades. But the grip of these conceptions of social change was so firm that they led one to believe that only the taking of state power could bring about a new society.

The late philosopher Henri Lefebvre was one of the rare figures to have anticipated some of the theoretical implications of the break with the old model. In 1978 he wrote:

> Between historicity and globalism a conflictual, that is, a dialectical, unity is unfolding. A third term is space and what occupies it, modifies it and transforms it into social space. Analysis must take into account the state and, of course, contending strategies, but also the distribution of productive forces and the division of labor on a planetary scale. The globalization of the state coincides with the extension and the reinforcement of the world market, and in particular with the entry into the market, alongside the most varied products and labor, of energy (such as oil and nuclear energy) and of information, of "gray matter," works of art, etc.[6]

But he took care to stress that "the state framework has a relation with logic and rationality, but it is not a relation of essence to essence or an abstract, unvarying principle of state." On the contrary, he called it an "instrumental relation."[7]

Michel Foucault, without suspecting the possible implications of his analyses, was not far from this conception of the state when he proposed at roughly the same time the notion of "governmentality." After pointing out that neither in the course of its history nor at the contemporary moment did the state ever have the unity, the individuality, or the functionality with which it was credited, he argued for abandoning a totalizing theory of the state and for alternative research on "the ordinary dimension of the state," that is, the procedures and acts by which the government of subjects and situations becomes operational. This meant studying its practices of adaptation, of attack and defense, its irregularities, its improvisations, in order to bring out other coherences, other regularities. Only a theory of concrete situations, a tactical analysis of the state and its techniques of government, seemed able for Foucault to open new ways of defining and redefining "what is the competence of the state and what is not, what is public and what is private, what is the state's domain and what is not."[8] This was precisely one of the major questions that in the 1970s was preoccupying the European ministers of culture.

Watching Over Culture

What is the relationship between industrially made cultural products and the access of the broad public to cultural goods? How are the socioeconomic conditions of creation modified by the possibility of multiplying copies of works of art and distributing them widely? Such were the enigmas which experts and advisers to the Council of Europe tried to resolve in October 1978 after observing that "the popularization of culture operates nowadays through the 'cultural industries.' "[9]

The European ministers responsible for cultural affairs, meeting the same month in Athens, went even further: "Given that a large part of the cultural industries are multinationals and that the means of exercising influence on them are complicated, we recommend studying the possibility of increased cooperation concerning international cultural industries—for example, carrying out a study of the degree to which the activities of the international or multinational cultural industries influence and are influenced by national cultural policies."[10]

It was indeed in the year 1978 that the concept of "cultural industries" made its grand entrance into the administrative language of a European Community body. Covering records, books, cinema, radio and television, the press, photography, artistic reproduction and advertising, and new audiovisual products and services, the concept was carried along by the new situation of competition between cultural policies traditionally carried out by the state, which reach only limited audiences, and the means of production and dissemination to a mass public, which are increasingly bound up with the international market. The very effectiveness of those policies was questioned. As it was put by Augustin Girard, head of the research department of the French Ministry of Culture and the man responsible for bringing the concept into the Council of Europe's frame of reference:

> By a curious combination of factors, some cultural policies, in their concern for democratization, have spectacularly misfired. State involvement in favor of the most deprived sectors of population living furthest away from large cities—an effort which has increased by 100, 200, or even 300 percent—has ended up working to the advantage of the most culturally favored and resulted in the expansion of central institutions to the point of ossification, while the people themselves have lost interest in public facilities, have

equipped their homes with cultural technologies and have been consuming the products of mass culture at home.[11]

To this crisis in the institutional management of "cultural democratization" must be added the imperative of defending national cultural identity against the avalanche of imported products:

> It would of course be absurd to speak of cultural self-sufficiency at the end of the twentieth century. Even if it were desirable, which it is not (cultures have always been transnational and have always nurtured one another), self-sufficiency would not be possible. . . . What one has to speak of, however, is cultural non-dependence, in other words a country's ability both to restrict superfluous imports and ensure competitive national production. Today only flourishing cultural industries well adapted to their environment can enable countries to take up this challenge.[12]

For public authorities, knowledge of these industries becomes a prior condition for the formulation of new forms of state intervention in this area. Understanding the functioning of "cultural industries" means closely analyzing the process of production/commercialization in its different phases of conception and creation, editing, promotion, distribution, and sale to consumers; uncovering the structures of industrial sectors (in particular, forms and degrees of concentration); and finally, determining business strategies.

A multidisciplinary team of French researchers had already undertaken such an analysis of the cultural products and services in 1975, laying the basis for an economic study of the cultural industries.[13] In pluralizing the concept of "culture industry," they intended to distance themselves from the postulates of the two Frankfurt School philosophers. For them a single culture industry did not exist as such; it is a composite collection, made up of elements that clearly do not belong to the same field, or at least are strongly differentiated. Analysis must bring out the factors that explain such diversity. This variety is found in the forms of labor organization (the role of the editor, the training grounds for more talent, the socioeconomic status of artists, etc.), in the nature of the products themselves and their content, in the modes of institutionalization of various cultural industries (public services, relation to the state, the role of the private sector), and again in the conditions of appropriation of the products and services by different groups of consumers or users. Moreover, capital does not make profits from cultural production without encountering resistances and limits.[14]

The undertaking in France of the economic analysis of cultural industries in the second half of the 1970s was contemporaneous, in Western Europe, with the development of what British author Nicholas Garnham called a "political economy of mass communications":[15] a vast theoretical and practical effort to escape the impasse of a critical theory still largely confined to first-generation semiological analysis.

Interdependence: The Diplomatic Change

Convinced of its singular position on the world chessboard, France tried from 1979 on to incorporate the notion of "cultural industries" into its new conception of foreign relations. This was the objective of an official report submitted in 1979 by Jacques Rigaud to the minister of foreign affairs. It was the first administrative document that postulated the necessity of more sustained linkages between culture, the economy, the diplomatic service, and private enterprise, within the overall design of a diplomacy in phase with the major trends in the industrialization of culture.

The report stressed the weaknesses of French cultural industries.

Along with politics and the economy, culture—in the broadest sense of the term—has become a component of international relations. . . . But it should not be disguised that, in the current state of the world, the products of the French cultural industries suffer from manifestly inadequate distribution, which results from a lack of adaptation, not to say an obvious Malthusianism. . . . Unfortunately, the cultural industries are oriented to the internal market to an exaggerated degree. Dispersed, badly organized, with weak investment capacity, they are in some ways marginalized and unknown, too commercial for their cultural attributes and too cultural for their commercial attributes. . . . The dissemination of French cultural products throughout the world remains largely artisanal and archaic, which profits neither business or culture, and too often limits our influence to a small number of initiates who are familiar with the confidential distribution networks.[16]

For the first time, a report commissioned by public authorities stresses the obstacles posed by a certain conception of foreign service as public service. "The public service spirit . . . leads to an inadequate, incomplete and, in short, embarrassed consciousness of the economic consequences of our public action. We must dare to speak, in certain instances, of cultural trade and, more generally, of the economic results of cultural relations. Naiveté in this field is not a defensible attitude."[17]

Finally, this official document introduced into diplomatic language concepts that tried to redefine the very content of "international relations." In fact, Rigaud's report ratified the vocabulary of "interdependence" as updated by Zbigniew Brzezinski some ten years before.

> A country like France should, more than others, be sensitive to the notion of interdependence, because its culture has given and received so much. Since the Renaissance, she has shown the culture of others an attentive and respectful attitude. Our archaeologists, linguists, and ethnologists have greatly contributed to the knowledge of great civilizations the world over, and thus helped peoples who are their heirs to form their own cultural identity. Inversely, it is often with lavishness and disinterestedness that we have delivered to them a French culture which, by its humanism, was for them the means of access to the universal.[18]

This new semantics of international relations was conceivable only at the price of abandoning the notion of "dependence." This meant doing away verbally with the relations of domination and the long history of unequal exchange in the course of which the "world-system" was built. At a more operational level, in the field of international negotiations, this notion of interdependence tended to legitimate the subjection of national sovereignty to the principle of defense of common interests of the whole world community. Paradoxically, the concept of interdependence made its debut in major international bodies in 1974, when the U.N. General Assembly adopted the notion of a new international economic order. The recognition of "interdependence" was in reality the gage of a promise to let the Third World participate fully in the management of an international economy, and to participate also in a world that was ceasing to be bipolar (developed/developing) and was metamorphosing into a "multipolar globe."

Community Media?

UNESCO put the concept of "cultural industries" into circulation at nearly the same time as did the Council of Europe. But the concept carried too many indelible traces of its geopolitical area of production to be exportable unconditionally. In fact, it was especially in countries with a high involvement of the state in cultural policies that the concept penetrated into research and thinking pertinent to the conflictual relation between the economy and culture, between public culture and the practices of private actors. This explains the undeniable echo of the concept in Francophone Belgium and in Québec,[19] regions that combined a large government apparatus in the cultural field with a concern to defend a cultural identity they felt was in peril.

Although they did not have to be concerned about safeguarding a position of power in cultural geopolitics, French-speaking Belgium and Québec did have the benefit of being premonitory experiences. For a number of years, these two territories, which then had nearly the densest cable TV infrastructure in the world, had been deluged with their neighbors' broadcasting channels and thus represented laboratories for the internationalization of an audiovisual system. From this advanced observatory emerged a first general rule: only news programs produced locally resisted competition from foreign channels, while fiction was the most difficult front to defend.

But the special feature of these two vanguard situations resided in more than the competition between a public service and channels coming from outside. Québec first, and Belgium later, had inaugurated a process of decentralization of their audiovisual systems. Community radio and television stations in Québec served in the 1970s as a model for numerous groups fighting for a revision of the public monopoly of the airwaves in France, where "free radio stations" were criminalized. The evolution of "community media" showed how false were the dilemmas that were at the center of current debates. It indicated, first, that a public service maintained for corporatist reasons is as pernicious as a decentralization aiming only to deconcentrate power, and second, that to play the game of "media-as-ideal-tool-of-self-management" against "television-as-centralizing-image-of-concentric-power" does not necessarily result in enlarged possibilities for citizens' expression or the renewal of the public sphere.

The impossible quest for the "community" and "local" spheres as anchors for the creation of an identity and a space of so-called alternative media coverage served as a reminder that democracy is not necessarily where one thinks it is. *Small* is not necessarily *beautiful*. When the "local" is used to drive back the advances of the "worldwide" or the "international," one may find oneself excusing a movement that tends to diminish meaning and the capacity to act in concrete situations. The "local" is of no real interest except where it allows a better grasp, by virtue of proximity, of the interaction between the abstract and the concrete, between experience and the universal, between the individual and the collective.

How to avoid the "local" 's diminishing the possibility of understanding the broader reality from which the concrete takes its meaning? The most stimulating response concerning this articulation between the local and the world-space could already be found among certain journalists with long international experience. One of them, Claude Julien, then director of the monthly *Le Monde Diplomatique*, wrote in 1977:

All means of communication and information should strive toward a goal while always holding together the two ends of the chain: the "little" local fact and its faraway causes—scientific, financial, political, or economic. . . . The local media thus have, in any case, an irreplaceable role to play. But this role cannot be confined to the description of facts and local problems that the "main" organs of information ignore. It cannot treat the town or the neighborhood as an island cut off from the rest of the world: through the case of laid-off mill workers to whom it gives expression, it should retrace the thread linking their fate with the Egyptian peasants on the cotton plantations, with the laboratories where new synthetic fibers are developed, with financial maneuvers and international competition.[20]

This link between the local microsystem and deep tendencies of the world market and technology had already been understood by many farmers—often before many intellectuals—when they saw that the price of their crops was no longer determined in the financial markets located on the territory of their nation-state but rather in the grain markets of Chicago.

Two Conceptions of Policy

"Ten years ago," stated Ithiel de Sola Pool in 1974, "few communications practitioners thought in terms of any overall communications policy, and few communications researchers would have recognized policy research as an established category. All that has changed. Communications policy has emerged as a field of research."[21] The American sociologist defined this new field of study as "normative research about alternative ways of organizing and structuring society's communications system." One of the principal factors explaining this tendency resided in his view, in the "exponential growth in the rate of technological change." The examples he chose to illustrate the takeoff of research on policy formulation were all taken from the field of new communications technologies such as cable and satellite television, and telecommunications more generally.

If de Sola Pool subscribed to this research perspective, seeing in it a way of better enlightening decision makers, he nevertheless signaled its dangers. What he feared most of all was an overly fervent commitment of researchers to the "big questions of public policy." He even described as sins the "hubris of intellectual elitism" and the disdain of empirical analysis of facts to the benefit of "philosophical and conceptual-analytic deliberation of goals." This peril seemed to him all the more necessary to avoid since objections from industrial and professional circles could already be heard, to the effect that this research was "a euphemism for the restriction of their freedom." This was why he preferred not to lock research into the "archaic nineteenth century ideological argument about social vs. private ownership" and "national sovereignty." His formula was therefore a policy of small steps dictated by the logic of technological supply. This approach consisted mainly in asking about the economic stakes and the means of assuring the social penetration of new technologies of communications.

The issue indeed lent itself to polemics. In the circles of American university research, other voices subscribed to a logic of demand that, contrary to that advocated by scholars such as de Sola Pool, began from the analysis of individual needs and examined the capacity of new tools to respond to them. Choosing this point of entry meant questioning a great majority of existing empirical research, its priorities in the choice of subjects, and of geographical zones to be tackled. In this perspective, the evaluation made by Hamid Mowlana and Herbert Schiller—in direct response to those of de Sola Pool—merely confirmed the history we have retraced in the preceding chapters: the close relationship between the perspectives adopted in research and the needs of corporations and government.[22] At that time very few others stood up against such practices.

A veteran of studies of international communications, Ithiel de Sola Pool devoted a part of his article to the role of communications in development, and proposed making this theme an absolute priority in research on communication policy. For in fact this controversy among sociologists in the United States was only the domestic extension of an international debate on the theme of national communication policies.

In 1970 UNESCO, under the direction of Jean Maheu, had been commissioned by the U.N. General Assembly to "aid member countries to formulate a policy relative to the major means of information." Since its foundation, the international organization had worked on the formulation of educational, scientific, and cultural policies, but had never before ventured into this field. The first meeting of experts took place in Paris two years later. From this moment on, the consulted specialists took care to circumscribe the range of these policies: "Recognizing the great differences in the social and economic situations and the diversity of political systems in the world, the meeting has not tried to propose one single method, but to indicate the basic factors that must be considered in the concrete context of each country."

A series of intergovernmental conferences by region took up this theme in Africa, Asia, and Latin America. The most important was the one organized in 1976 by the Latin Americans in San José, Costa Rica,[23] out of which came a more precise definition of the concept:

> Communication policies constitute coherent ensembles of principles and norms designed to set the general orientation for communications bodies and institutions in each country. They furnish a frame of reference for the elaboration of national strategies with a view to setting up communication infrastructures which will have roles to play in the educational, social, cultural, and economic development of each country. Even when they are not explicitly formulated, national communication policies already exist in many countries; they represent the culmination of a process of cooperation and negotiation among various partners: public authorities, the media, professional bodies and the public, which is the final user.[24]

The theme of the "national communication policy" had already been in the air for a number of years, particularly in countries such as Venezuela, where the Christian Democratic and Social Democratic parties governed in turn and where scholars and representatives of civil society had already elaborated together the principles of a policy for their country.[25] The objective of the participants in the Costa Rica conference was to find policy formulas that put citizens on guard against both the arbitrary power of the state and the shortsightedness of the market. They aimed especially to introduce, into audiovisual landscapes given over to an unchecked market logic, a corrective inspired by the philosophy of "public service."

Opposition from local and regional business circles was not long in coming. The same lobbies that had grown up to oppose the principle of a new international order in information lent them support, interpreting this proposal for public control as just one more stratagem to assure the stranglehold of the state on freedom of expression. Very few states took the initiative of promoting these principles, the gist of which was simply to make all parties admit that communication was not a commercial activity like all others. Proposals to involve citizens in deliberations on the place of the media in social development, in particular through representative councils reflecting national plurality, were shelved, even in countries where governmental authorities had been the most vigorous proponents of democratic restructuring of the means of communication. As for authoritarian regimes that had been in power for years, or even decades, as in Brazil, South Korea, or Indonesia, they had had no need to wait for the concept of "communication policy" to be born to perfect a policy of censorship affecting information and the media as well as all forms of public expression by their citizens, in the name of national security or the defense of public morality.

The Ambiguities of Expertise

Independently of their procedural implications, the debates on the regulatory role of the state demonstrated that there was another way of linking research, expertise, and policy formulation than the one imagined by de Sola Pool, which was complicit with the status quo.

Many scholars actively engaged in the policy of democratizing communication dared to raise in their own countries those questions of a "philosophical" nature which so frightened de Sola Pool. For them, one of the essential problems posed by policymaking was precisely the lack of autonomy of existing research bodies, which made it difficult to carry out research that strayed beyond the beaten paths of bipolar schemas, and which did not tend to legitimate the positions of the state or the private sector. As the Venezuelan Oswaldo Capriles put it:

> In the Third World there exist few research institutions that provide *sufficient autonomy*: lack of economic resources, lack of sufficiently trained human resources, political dependence, and in some cases, of certain institutions with respect to the government, and even the censorship or ideological control of the research; imposing influence on the topics on the part of foundations, governments and institutions sponsoring the research. In many countries, predominance of research for the private commercial sector: marketing, opinion studies, ratings, advertising, etc. . . . Committed research thus appears or tends to appear as an exception. A frequent campaign in dependent countries tends to undervalue the university research worker before public opinion: he is either a political extremist, or a utopian thinker out of touch with reality.[26]

The issue recurs regularly and its origin goes back to 1941, when Paul Lazarsfeld had formulated the dilemma of administrative research versus critical research, clearly taking his distance from the second alternative.[27] This dilemma had long masked the theoretical poverty of empiricist sociology, a sociology that Theodor Adorno had justly considered to be incapable of "epistemological distance" as a result of having reduced the notion of "methodology" to "technical practices of research." The legitimation of "administrative research" was what allowed Ithiel de Sola Pool to avoid having scruples about his promiscuous relations with his country's national security policy even while denouncing others in the name of the scientist's sacrosanct neutrality.

At the end of the 1970s, questions about this line of demarcation were more than ever on the agenda, and not just in the Third World. The report drafted in 1977 by Elihu Katz at the request of the BBC was a case in point.[28] The British television network had solicited the expertise of this sociologist to map out the priorities of its research policy.

The Katz report marked a turning point in expert research in the service of an institutional television apparatus. It was interpreted in this light by British and North American scholars such as James Halloran and James W. Carey, who did not fail to subject it to critical evaluation. Roughly speaking, the report did no more than recycle the tired theme of media "effects" on audiences, without ever asking how it is decided what appears on the TV screen.[29] Katz never risked a more explicit vision of the relation between television and society, television and social stratification, television and group psychology, or television and culture, instead engaging in intellectual shadow theater. Nor did he ever consider that the history and the intentions of the observer are an integral part of the history and the meaning of what he or she observes.[30] So there is nothing surprising in the implacable verdict of his colleagues: "The proposals were developed to meet criteria stated in advance by the BBC. . . . By accepting the policy framework, the Katz report is not an encouragement to 're-think' television or a call for scholarship about it—it is a call for research on it. That call must pretty much assume the existing social structure and political arrangements and the existing role television plays in our personal, political, and cultural life. We are back in the bind of administrative and critical research."[31] And it is this function of legitimating decisions already taken that disturbed all those who were aware of the quantitative and qualitative jump that expert research was then taking.

The Crisis of the Federating State

Is industrial strategy in the area of information and communication technologies an essential pillar of "communication policy"? Rare are the public authorities throughout the world that have raised this question or allowed themselves to formulate it. If France at the beginning of the 1980s was able to attempt to resolve it, it was because the state had always been a majority partner in the electronic industries. In 1978, the report by Simon Nora and Alain Minc had pointed to information and communication technologies as a way of "overcoming the crisis" and had proposed a voluntarist state strategy. Yet it was only in 1981, after the first election of François Mitterrand, that the French government elaborated the principles of an industrial policy in harmony with that objective. This was the role incumbent on the "integrated electronics sector" (*filière*).

To guarantee a future of technological independence, Nora and Minc had proposed choosing a policy of product and market niches. The new public authorities were in fact to begin with another postulate, that of the interdependence of various electronics subsectors, and therefore, the impossibility of backing only one or even several of them, and hence the necessity of investing in them all.[32] The concept of "integrated electronics sector" (*filière*) designated the process of unification that linked together eleven separate areas of the electronics industry (components, consumer hardware, computing, office systems, software and data banks, the automation of production, medical electronic and scientific instruments, telecommunications, professional electronics, electronics for weapons, and space technologies). While each of these sectors had its own logic, none could be envisioned without this progressive development of interdependences.

This diagnosis concerning technological convergence to be carried forward by digitalization and the homogenization of components controlled the reorganization of the whole network and imposed a lifting of barriers between sectors. The federating role of the state proved to be determining. In fact, the French strategy drew some lessons from the organizational experience of Japan, whose industrial success owed much to the incitement of the Ministry of International Trade and Industry (MITI). Overcoming barriers between subsectors meant establishing a new scheme of relations between the public sector and private enterprise, between industries in the same branch (in order to avoid competition *within* France), between national firms and subsidiaries of multinationals. Finally, it implied rethinking strategies of innovation by assuring the fluidity of horizontal transfers of technologies: strengthening the ties between industry and universities and between teams of private and public researchers; and integrating the design and marketing of the product by involving potential users in the processes of innovation.

This strategy involved uncertainties, including limited financial resources, the time necessary to guarantee a growth in foreign trade balances, and the uncertainty of social and cultural acceptance of new technologies. A necessary condition for success was to concentrate efforts at the European level in order to pool resources and broaden the scope of negotiations concerning the conditions of competition with major players such as Japan and the United States.[33] As it turned out, the economic environment did not live up to expectations. In ten years, the commercial deficit of the European electronics industry grew twentyfold.[34]

In addition to the difficulties of building a European economic, cultural, and social space, the challenge of competition from Japanese firms, and the impossibility of combining Dutch, British, German, and French firms, the final blow for this "integrated electronics sector" came from the measures taken by the neoliberal government that came to power in France in 1986, barely four years after the official ratification of the concept: denationalization and the end of the federating role of the state.[35]

Eric Le Boucher, one of the initiators of the notion of integrated electronics sectors, made this disillusioned observation:

> The Nora-Minc report was written at a time when the values of the left in France occupied the ideological foreground. It attempted to launch a broad, political vision of things to come. Conditions were no longer the same a few years later. "Pragmatism" was to squeeze out reflection. On the left the schema became the same as on the right: the more the country is computerized, the more it "modernizes" and the greater its chances of "winning out" against others in the crisis. . . . To say that "the computer is only a tool," as was hammered home by the adulators of "modernity," comes down to deliberately rejecting any thought about the tool itself . . . in short, to approving without reservation the spread of the cold medium of a modernity without a vision.[36]

Deregulation and globalization now had wind in their sails. "Modernization" was the horizon. It was an idea that despite the accumulated criticism of the preceding decades sociologist Alain Touraine—to whom we owe the importation into France, in the late 1960s, of the concept of the "postindustrial society"—hastened to acclaim as liberating:[37]

> French society is no longer made up of social classes: it is divided into three great categories, defined by their relation to modernization. . . . We find ourselves delivered from all visions of future society, from everything that hid an experienced reality that no longer reached us except through the voices of a few singers and clowns. Politics has finally been deconstructed. We are in the era of post-liberalism as well as postsocialism.[38]

With the advent of visionless modernity, governmental discourse lost a whole range of references introduced into the state's representation of international communication in the brief period when it still seemed possible to redefine the balance of world power. That period had featured in 1982 both the Cancún conference on North-South relations and the international conference in Mexico on cultural policy, where the French minister of culture, Jack Lang, provoked the wrath of the U.S. delegation because of a speech in which he lambasted "cultural imperialism."

In the spirit of the first years of the socialist government, President

François Mitterrand presented a report entitled *Technology, Employment and Growth* to the summit of the industrialized countries at Versailles in June 1982. He roundly denounced the threat posed to "memory" and the "freedom of thought and decision" by the financial, industrial, and geographical concentration of information in the hands of a few dominant countries, and he sounded the alarm of the risk of "uniformization of cultures and languages." Appropriating the themes and concepts of the "new world order of information," "codevelopment," and the transfer of technology, Mitterrand proposed a World Charter of Communication with a view to "guaranteeing to countries of the South the capacity to control their means of communications and the messages they carry" and "protecting the sovereignty and cultural integrity of states, menaced by new technologies."[39] In short, this approach, which dared to question a certain growth model and its logics of exclusion, contrasted sharply with that of his partners in the club of seven industrial powers, which were not yet the "board of directors of the world" they were to become less than ten years later.

Chapter 10

The Ascendancy of Geoeconomy: The Quest for Global Culture

The "Economic War"

In his "Letter to the French" of April 1988—which was intended to offer a program for his second seven-year term as president—François Mitterrand wrote: "Let us consider the world economy; it appears as nothing but a battlefield where businesses wage a pitiless war. No prisoners are taken. Whoever falls, dies. As in military strategy, the victor always follows simple rules: have the best preparation, the fastest moves, take the offensive on adverse territory, have good allies and the will to win." This martial discourse reflects the spirit of the 1980s.[1] "Clausewitz to the rescue of marketing," "the battle for spoils," "officer of the new world conflict," "psychological movements of images on demand": these are a few expressions gleaned from strategic treatises and financial magazines, examples among others of the combat noises heard among corporations seeking to continue politics by other means in the theater of world operations. Needless to say, the semantic aggressiveness of the new officers and high priests of the economic war would appear derisory and shameless when real war broke out in the Gulf.

Throughout the world, corporate actors had grown in visibility, not only in the great financial markets but on the best-seller lists. For his first book, Thomas J. Peters, American guru of "excellence," pocketed nearly two million dollars in royalties.[2] In France, the largest publisher specializing in this type of work was publishing a hundred books a year in 1989, as compared with ten during the preceding decade. These books, which enjoyed a transnational readership far broader than just business executives, provided a medium for the followers of the new business doctrine—a doctrine relayed on the French domestic market by the new business supplements of the large newspapers and new magazines with such evocative titles as *Défis* (Challenges), *Challenge, Dynasteurs,* and so on.

Playing on words, one could say that a transition had occurred from the hegemony of the reason of state (*raison d'état*) to the supremacy of corporate status (*raison sociale*). The norms and references of the welfare state, public service, and the constraining play of social forces, all tended to cede power to private interests and the free play of market forces. Not that the state tends to disappear or lose its monopoly of rule-making, nor that the two "regimes of truth" are impermeable one to another—far from it—but the fact that the corporation and the freedom of the entrepreneur became the center of gravity of society resulted in a redistribution of hierarchies and priorities and the roles of other actors. In short, what changed was the whole way of producing consensus, of cementing the general will.

"The field of management," wrote the sociologist Michel Vilette in 1988, "has contaminated all segments of society and is perceived as a universal cultural model."[3] Not only did the corporation become a full actor in public life, expressing itself more and more openly and acting politically on all of society's problems, but its rules of functioning, its scale of values, and its ways of communicating all progressively impregnated the whole body social. "Managerial" logic was instituted as the norm for managing social relations. The state and its territorial subdivisions as well as associations were seized by the schemas of communication already tried out by the protagonists of the market. The portfolio of offers of professional communication services was enriched by new clients and new skills, and the very definition of communication gained a new set of conceptual frameworks.

This process of the social emergence of the entrepreneurial actor both fuels and is fueled by deregulation. By this term I mean something that goes beyond the liberalization or lifting of rules and laws that had bridled free enterprise, and well beyond the "fluidification" of circuits of finance, transport, telecommunications, and audiovisual media. The process of deregulation can only be interpreted as the promotion of a different principle of social organization, another way of putting individuals, groups, societies, and nation-states into relation with each other. It is also another way of doing (or not doing) social theory.

This new way of thinking involves not only free enterprise but also a veritable cult of enterprise, bordering on the religious, to the point where many firms have taken their desires for reality, their project of corporate development for internal democracy, their discourse about new internal communications, for the advent of employee participation and mobilizations, and new forms of corporate self-organization for new means of personal realization.

With the consecration of the entrepreneurial approach, other voices in the knowledge and practice of communication began to be heard. If during the 1960s and 1970s the state and its major officials, as well as academic circles, were the nearly exclusive producers of discourse on communication, proposing concepts and protocols of action, in the 1980s, on the other hand, market experts progressively took over the field. Even if complicity between the university and commercial expertise does not date from this decade, it is a euphemism to say that the passageways between the two have multiplied quickly, and that even beyond this type of exchange, more and more notions circulate about communication that are in fact put forward by corporations themselves or their networks. Moreover, one may reasonably foresee that this phenomenon of telescoping together "practical knowledge" and "theoretical knowledge"—if indeed that distinction still really means anything—can only grow. The liberation of flows signifies the acceleration of the circulation of knowledge, including theoretical knowledge, in all directions.

This multiplication of crossroads and meeting places often happens to the detriment of meaning and proper epistemological distance, as certain contemporary efforts to theorize managerial action eloquently show. With the intellectual legitimacy conferred by innumerable references to Jacques Derrida, Michel Foucault, and Jean-François Lyotard, such works claim to explain to us the birth of "postmodern enterprise." The enterprise of the 1980s becomes an immaterial entity, an abstract figure, a universe of forms, symbols, and communication flows, in which the problems posed by the restructuring of the world economy and the redistribution of the dependencies and hierarchies on the planet become diluted.[4] No more traces exist here of the celebrated model of the new apparatus of control, that Panoptikon of which Foucault made himself the archaeologist, but instead a vaporous world of flows, fluids, and communicating vessels evolving into "dissipative structures."[5]

The Global Dimension: A Mode of Management

The space of organizing production and marketing has extended itself to the space of the world market. "The global" and "globalization": these two notions—directly transposed from English—have taken over from "international" and "internationalization."

The new stage of expansion of corporations with a world scope comes in the wake of four previous stages. At first a firm was content to export through a network of distributors and local suppliers. Then it constructed its own marketing network. Later, while remaining essentially based in the country of its headquarters, it started to establish its own factories and its own sales and marketing networks within certain markets and in foreign economies that it saw as key. Finally, it integrated itself into national industrial contexts, sometimes even agreeing to decentralize its research and development centers. This history was, however, less linear than it appeared. Different stages were completed at varying rhythms, or even skipped altogether by different branches of industry and services. This history of the international expansion of corporations is therefore asynchronic and will continue to be so for a long time, to the extent that the global vocation is not the same in each corporation. But the industrial and commercial corporations at the forefront of the new mode of internationalization clearly constitute the standards of excellence for the others.

One of the prerequisites of the transition is the gigantic redeployment marked by leveraged buyouts, transnational alliances, and the megamergers of the 1980s, guided as they were by three slogans of the decade: economies of scale (how to produce most cheaply); the power of scale (how best to manage thanks to the accumulation of networks, information systems, and talents); and economics of breadth (how to cut costs by producing several different products within the same line, or diversification within standardization).

In the media and advertising sector, this redeployment helped constitute the phalanx of new planetary networks of agencies—basically American, British, French, and Japanese—as well as new multimedia groups, in Western Europe, Australia, and Japan in particular. By crossing the Atlantic and bidding to take over U.S. firms, these new arrivals provoked the megamerger of the century: the alliance between the U.S. giants Time and Warner. Communications appeared a highly profitable sector, and the stock market helped it on its meteoric rise, until the day when, for all parties concerned, a conjunctural decline and fluctuations in the advertising market made these huge publicity networks and media corporations much more fragile.

The 1980s, which had seen the rise of giant-scale ventures in an industry doped up by its own talk about exponential growth, ended in uncertainty for investors, and for banks in particular. As for the effects of "industrial synergies" and the power of scale, the strategies for multimedia or multiservice diversification sensationally announced by candidates for "global communications corporations" or complete horizontal and vertical integration failed to produce the expected results. The press magnate Robert Maxwell—a few months before his death and the collapse of his group—divested his audiovisual activities, and the "total" communications agency Saatchi and Saatchi had to renounce merging into a single portfolio of services management consultancy and advertising expertise. These instances remind one that building a global corporation is more chaotic than one might have believed during the feverish bidding of a leveraged buyout.[6]

The semantic transmutation from "international" to "global" took place so rapidly that theorization is overwhelmed by professions of faith. Nor is it at all likely that theory can catch up, considering the pressure of pragmatism. This is one reason why it is difficult to separate the theoretical analyses proposed by management specialists, for example, and the discourse of major industrial firms striving to reach critical mass or advertising agencies seeking to position themselves in a transnational market. That is also why it is difficult to separate the triumphalist myth of immediate, gung-ho globalization from its fragmented and chaotic realization. This gap never appeared so clearly as when a heavyweight of management science, Professor Theodor Levitt, director of the *Harvard Business Review*, justified by his writings and counsel the corporate strategy of the British-based global communications agency Saatchi and Saatchi. This strategy collapsed in less than three years under a colossal indebtedness and the pressure of the recession on the U.S. and British advertising markets. It was this very same Harvard professor who had theorized the concepts of "global" and "globalization," though he had not invented them, since the terms were already circulating in the marketing divisions of major world firms.

The concept of globalization is equally indebted to the matrix of "financial globalization" that developed in the course of the 1970s and 1980s, a period when the framework of the financial systems in place since the end of World War II broke apart and frontiers between traditionally different occupations and national systems dissolved. New products and markets appeared in the financial sphere; all were immediately international in a world economy of real time.

The Amoeba

Global space gives rise to the notion of total management, or better, "holistic" management, a term common in English though rare in French except in philosophical dictionaries. It refers to "holism," the theory that the whole is greater than the sum of its parts.

Levitt's approach is based on four observations: the world is becoming a "global village"; the market is no longer national but worldwide in scale; the urban way of life predominates; and certain major tendencies can be observed (the development of individualism, the Americanization of youth, the emancipation of senior citizens, and so forth). From these observations spring three hypotheses: the homogenization of needs under pressure of new technologies; price competition (consumers are ready to sacrifice their specific preferences to take advantage of cheap and reasonably good quality products); and economy of scale (the standardization made possible by the homogenization of world markets permits the reduction of costs). Levitt recommended that corporations create a single product for the whole world market, market it at a single price (the lowest possible), promote it in the same way in each country and use the same distribution circuits everywhere. In other words, roughly, he recommends that they imitate firms like Coca-Cola, one of the few that follow such a strategy.[7] This theory of the homogenization of needs and markets and the standardization of products has been subjected to numerous criticisms by those who think that the world, on the contrary, is becoming more differentiated and who recommend a return to the original definition of the term "marketing," which is to segment the market according to the differences that run through it.

Globalization is both an internal and an external affair. It is a way of organizing a firm and a way of relating to world space. To describe this new mode of organization, the economic literature calls on the metaphors of the hologram, the amoeba, and frequently the language of biology. The corporation and the world-as-market are treated through the prism of the living organism.

We must understand these metaphors as signifying an end to the rigidity of hierarchies within the corporation, the decline of forms of pyramidal authority inherited from the military conceptions of managers shaped by World War II, in which the retention of information was a source of knowledge-as-power and where everything functioned by sanction and penalty. By contrast, here is a model of management based on information and communication networks, in which personnel are implicated and made to feel responsible for fixing and realizing objectives, and in which positive criticism seeks the harmony of networks of interaction, tapping employees' informal and spontaneous creativity and capacity to innovate. This model involves appropriating knowledge and skill and reinvesting them continually in the organization. As the director of Gannett, one of the most dynamic press groups in the United States, put it: "I want each of my journalists to become his own manager and marketer." In the flowering of neologisms that accompanied the boom in best-sellers on management, one notion sums up the change: the "intra-preneur." To Fordism's separation of tasks is opposed a new form: the capillarity of the managerial function, its diffusion within the body of the enterprise.

And since the employee is a part of the whole, he or she is also a carrier of the whole.

Globality as a mode of management has no meaning unless it is linked to the corporation's mode of inserting the business into the world economy and world market. There, too, the corporation obeys the relational schema. Management of production, marketing, and research and development operates in a mesh made up of all the internal networks of the firm, interconnected with external networks. This is demonstrated by corporations' share in transborder data flow: 90 percent, according to a 1987 estimate.[8] Corporations depend on these flows at five levels: production coordination planning systems, including production scheduling, management of production materials, and inventory record-keeping; financial systems, including budget monitoring and accounting systems as well as sales and production data; engineering systems, including large-scale mathematical models and specialized computer equipment; purchasing and customer systems, including accounts payable and receivable, historical profiles of suppliers, corporate customers and airline reservations; personnel management systems, including payroll and human resources planning. Fed by an incessant flow of information, the network-corporation abandons its vertical and centralized structure and adopts fluctuating contours. Consider, for example, the more and more frequent call for subcontracting (which may give rise to further subcontracting). The network transforms and regenerates itself constantly.

This mode of organization places the corporation in the first rank of clients for integrated communication services (radio, television, visionphone, voice messaging, data transfer, telecopying, etc.), which open the way for the unification of systems across normalized networks such as ISDN (integrated services by digital network). A portable terminal allows each subscriber to receive polymorphous messages transmitted to him or her from any point on the planet connected to the network.

Users as Coproducers

To the hierarchical distribution of tasks and powers in the Taylorist enterprise corresponded a sedimentation of space. The local, national, and international represented tiers, impenetrable one to another. The new representational scheme of the corporation-as-world-network proposes a model of interaction among these three levels. Every strategy on the world market must be both local and global. This is what the Japanese managers express with their neologism "glocalize."

The first element was the creation of an adequate corporate culture. The horizontal model of organization presupposes generating or reinforcing in employees a feeling of belonging to their company. In the world-scale firm, corporate culture, as a sharing of values, beliefs, rituals, and goals, has the mission of forging the impossible alliance between the local and the global. This culture is not properly speaking situated in a territory: it is a mentality. Once this goal has been enunciated, the new breed of manager is rarely more precise, providing at best a few recommendations: do not let national identity have the upper hand over global identity, recruit native managers, promote them to positions of responsibility in international management, develop multicountry career paths, do not penalize expatriates, multiply the opportunities for trips abroad, and finally, reaffirm, in a coherent corporate policy, the objectives of the firm's global strategy.[9] A level absent from this mystique of corporate identity and culture at the vanguard of the world economy is that of the individual and family. The time scale of global management completely destructures the notion of free time, with unfortunate consequences for family life and employees' nervous systems. As for the excluded categories—"stateless" victims of the new corporate patriotism, products of the generalization of precarious labor—they do not react favorably to incitements to identify completely with their corporation.

A second aspect of the new configuration is the decentralization of certain decisions. While more than ever the center makes global managerial decisions on strategic issues, products, capital, and research, and while more than ever the headquarters is the summit of the network and the heart of information-gathering and distribution, a certain "decentralization of the center" to the advantage of local units is nonetheless called for. Their power consists in deciding tactical questions such as marketing, packaging, and advertising, or else deciding on modulations of a standardized "core" product.

The role attributed to advertising and marketing in establishing a link between the local and the global is indeed a decisive point in the strategy of globalization. As Kenichi Ohmae, director of the Japanese subsidiary of McKinsey Consultants, explains: "[The] value system [of the global corporation] is universal, not dominated by the dogma of the home country and it applies everywhere. In an information-linked world where consumers, no matter where they live, know which products are the best and the cheapest, the power to choose or refuse lies in the hands of customers, not in the back pockets of sleepy, privileged monopolies like the multinationals of an earlier time."[10]

According to the Japanese expert, this return to the consumer proves all the more necessary in that this proximity acts as a counterweight to the "centrifugal forces" that push managerial actions and decisions far from the consumers' sphere of control, far from their orbit. The more the centrifugal logic is felt, "the more dispersed your people and the closer the attention they must pay to local customers and markets, the more they need to escape the center's rigidities while retaining its shaping values."[11]

The supplier tends more and more to search upstream to discover the client's desire and even to help him or her define it. As economist Albert Bressand notes: "From production pure and simple, we are moving to 'coproduction': the value added to the product is found more and more in its fine tuning to demand."[12] This move to the consumer raised to the rank of "coproducer"—or "pro-sumer," in Alvin Toffler's expression— finds a powerful auxiliary in the gathering, storage, and treatment of information. Consider private industry's restructuring of research and the prodigious development of multivariable studies of merchandise flows, program flows, and the flows of audiences, which began with the first applications of scanning techniques in supermarkets (bar-coding). This trend is also seen in data banks and data bases that combine more and more variables to identify and classify targeted groups according to common characteristics. This evolution doubtless reflects the growing "Taylorization" of consumption. More and more disciplines, more and more people, are endeavoring to scrutinize consumers' acts and gestures for strategic purposes. As a result, administrative research has undergone a qualitative and quantitative leap in countries that had been more or less spared from market logics until recently.

Standard Supply or A la Carte?

The most revealing issues on the frontier of homogenization have crystallized around the defining of targets for advertising campaigns seeking to respond to the new exigencies of the world-as-market (*marché-monde*) or its regional subdivisions.

The space occupied by the advertising industry and by marketing has grown to the point of their becoming an unavoidable mediator of the interface with the media. Agencies and groups have made themselves global in a geostrategic sense—with certain advances and retreats, of course —pushing the ramifications of their transnational networks ever outward. The integration of services offered has advanced along parallel lines. The failure of Saatchi should not allow us to overlook the logic of diversification of advertising activities extending well beyond media advertising to embrace sectors as diverse as direct marketing, sponsoring, institutional communication, and lobbying. This tendency has reached the point where advertising conceived for managerial purposes tends more and more to be dubbed "communication," further complicating the definition of the field. Deregulation of audiovisual systems has brought its share of novelty to the advertising scene by bringing together the large agencies and advertisers in the sponsoring and producing of programs.

It is against this background that the different efforts to determine the profile of the transfrontier consumer have been pursued. The hunt for cultural universals is on. It draws on investments already made by mass culture in the imaginary (*l'imaginaire*) of people belonging to very different cultures. The creation of a single market of images is one of the stakes in the redeployment of the audiovisual industry. But the vicissitudes of the construction of a Pan-European television industry already suggest the difficulties in bringing together for one single program at a given hour an audience composed of a mosaic of cultures and languages. Channels of this type have already had to lower their ambitions. The older story of the internationalization of the daily press and periodicals has long since suggested that internationalization is a matter of segmentation and asynchronism. For example, a given magazine is not internationalized everywhere at the same time (there is a vanguard as well as a rearguard); a magazine is not necessarily adaptable to every country; the target aimed at by local versions of international magazines is mainly the upper sectors of the "middle class."

The experience of advertising agencies tells the same story. It demonstrates that, aside from products already hallowed on the world scene, such as Marlboro, Coca-Cola, or Levi Strauss, there are often more differences than similarities. A congress of the IAM (International Association of Marketing), held in Paris in April 1989, ended on the following observation: there is indeed a profound tendency toward globalization of markets and economies, but it coexists with other tendencies leading to the "generalized demassification" of consumption and the blossoming of micromarkets, a logic just as tangible as the simultaneous internationalization of macromarkets for consumer durables. A more nuanced approach is thus called for, to take into account the differentiation of consumer tastes, with an à-la-carte offer gradually taking the place of a standard supply—at least in the major industrial societies (since the debates did not venture beyond this perimeter). This perspective of "differentiation within globalization" stimulates the search for transnational segments, as revealed by studies that track differences and similarities beyond national borders in the attempt to circumscribe sociocultural mentalities—that is, the major groupings of individuals who share life conditions, value systems, priorities, tastes, and norms.

And so the notion of "Euro-styles" is born. While Europe still desperately seeks a unified identity, this notion proposes a new sociological breakdown of the population of the old continent. This provides an insight into the prevailing effervescence concerning the idea of a single European market—but also of the vagueness of the "expert advice" provided by agencies that are more interested in beating their competitors than in submitting to scientific verification their hypotheses on particular identities, differences, and resemblances.

It also shows that one cannot explain the redistribution of the spaces of transnational communications without taking into account the dynamics of the creation of vast supranational zones of commercial exchange: not only the laboratory of the European Economic Community with its twelve member states but also that of the North American Free Trade Agreement (NAFTA) between Canada, the United States, and Mexico, not to mention the pole now forming in Asia around Japan.

The European testing ground has already changed the strategies of large corporations with worldwide ambitions. One of the central issues at stake is the necessity of finding a proper balance between internationalization and regionalization, at the level of organizing principles and structures as well as that of products and strategies. This is what IBM calls the "challenge of transnationality," appropriating a notion born in the 1970s in the United Nations expert commissions responsible for formulating codes of conduct for the multinationals. The first timid commitment by IBM was its decision to transfer the headquarters of one of its six divisions from the United States to Great Britain. The production and development strategies of each of the six divisions, however, continue to be defined in the United States.[13]

The slowness of the process of building "transnationality" at the level of the production system makes it all the more necessary to build it by means of symbols. It is in this sense that the consumer is the essential agent of legitimation of the globalization of the economy, although the discourse concerning the simultaneously global and differentiated consumers is still full of holes.

Exclusions

As a paradigm not only of the corporation but also of future society, the conception of globalization calls for three observations.

First, the question needs to be reframed somewhat. The corporation, conceived as a communication and information network radiating outward from itself, is a choice field for applying systems theory. This theoretical corpus, which unifies the most diverse forms of communication and information around notions such as "system," "autonomy," "complexity," and so forth, is doubtless rich in potential for understanding the complex dynamics of our social systems, because it is willing to cross-fertilize both new and traditional scientific fields (molecular biology and evolution theory, neurosciences and artificial intelligence).

But it is still necessary to take precautions, so tenacious is the ideological heritage in communications studies of that old positivist project, a sociology of the "social organism." Systems theory is indeed potentially dangerous in the uses to which it is put, as one biologist explains: "It is with apprehension that one sees some notions—which seem to confer the status of prophets on the exponents of the "new paradigm"—become perverted and changed into a fashionable jargon or an ideology of the third millennium, when they are appropriated by business or media circles."[14] Every social problem tends to be formulated as a communications equation. If there is a difficulty, a knotty problem, it is reduced to the loss of a piece of information (due to filtering, a blocked channel, entropy, etc.). This vision of the corporation is potentially dangerous because it risks justifying the great imbalances and inequalities that divide the planet. In particular, its theories of self-organization, while they have the virtue of attenuating the idea of a pyramidal organization of power, fail to take into account their necessary obverse—that is, the exclusion of everything external to the system of reference, or what the same biologist quoted above calls the "exo-organization." But this external world is full of sound and fury, contradictions and conflicts. These potential neo-Darwinian deviations are all the more worrisome in that this conception of the corporation sees itself as a new conception of the universal. In reality, this universal is one that functions in a closed circuit. It is no coincidence that Kenichi Ohmae, the author who has pushed the global concept the furthest, is also the one who has frozen the world of the global economy into the notion of "triad power," that is, the area occupied by North America, Western Europe, Japan, and the new industrial countries of Asia.[15] The theory is clearly segregative, for it considers the remaining 80 percent of the world population only as candidates for the model of consumption and the way of life prevailing in the triad. There is a déjà-vu quality about this idea, since Ohmae is only renovating the old diffusionist axiom. In the same way, deregulation is merely the continuation by other means of the old principle of *the free flow of information.*

Guided by this principle, the notion of self-organization paired in the 1980s with that of self-regulation. The latter has become the natural complement of the project of deregulation.[16] Thus, during the recent European debates on the harmonization of legislation on advertising in the audiovisual media, it was in the name of the right to self-regulation that interprofessional organizations (grouping together advertising clients, agencies, and media) asked the European Community to raise the statute of "free commercial speech" to the status of a "new human right." What this actually meant was the freedom to push back the limits imposed by society on subordinating the public sphere to the ends of publicity, as Jürgen Habermas would put it.[17]

The second remark reinforces the preceding one. It applies in particular to a conception of the corporation such as that developed by Ohmae. For if this biomorphic theory of corporate organization is cosmopolitan, it also reflects the characteristics of the culture that enunciates it. Reading the works of this theoretician of management, one cannot help being struck by his conception of what one could call the welfare corporation. It is to anthropologists that we should look for the beginning of an explanation for this. Japanese society tends to assimilate bonds of kinship and local ties and, more generally, to identify the social with the biological, culture with nature. The language itself reflects the analogy between family and business. The traditional terms *oyabun* (both father and boss) and *kobun* (both child and employee) express it directly. "Groupism," which pushes the members of a group to serve the interests of that group, has been institutionalized in the large modern enterprise; it is the transposition and conscious adaptation of a model drawing both on tradition and on nontraditional realities in which a need for regulation makes itself felt. Hence the strong identification of the worker with the enterprise. As for the deviant, who does not respect the rules of the group, he or she is excluded. But "groupism" means that deviance is rare.[18]

The third remark concerns the handling of crisis, that is, all the events that might destabilize the "reference system." The corporation is only viable under two conditions: It must be communicative and integrated. The social and economic environment, however, is more and more unstable and hostile. Hostile takeovers, ecological catastrophes, labor conflicts, mergers, restructurings, terrorist threats, product sabotage, and other accidents and incidents are hypothetical situations where the "crisis" takes form and demands an immediate response. The spilling of barrels of toxic substances into a river, the sinking of a supertanker, an accident in a nuclear generator, the explosion of chemicals—such catastrophes are manifestations of what have been dubbed "major technological risks"; they forced the corporation in the 1980s to study crisis management.

Crisis is not only the heated moment in which the corporation must mobilize its networks to attenuate the impact of dysfunctional elements on public opinion and on its employees. Crisis becomes an integral part of the way of managing the corporation in peacetime. As the communications director of the Sandoz corporation put it during the 1986 catastrophe, when toxic products flowed into the Rhine after a fire in a Basel warehouse: "We must develop a cybernetic perception of public relations. . . . To get there, corporations have the duty to reflect in times of peace on their culture, their ethics of communication, and the choice of their communicators."[19] In the corporation conceived as an extremely complex interactive system that must face the irrational within and the unforeseeable without, the circulation of communication flows must not be interrupted. Information is life itself, a vital flow for staying in tune with the times, hence the permanent conflict between the need for transparency and the maintenance of an image. Hence, as well, the difficulty of going beyond a transparency understood as anything other than the struggle for the legitimacy and credibility of the enterprise by means of communication. But transparency, if one were inclined to dreaming aloud, could turn into a vast questioning of the prevailing model of development and progress and its natural propensity to destroy the environment—a questioning very different from the ecological marketing of the "green" political line.

The Rise of Japan

It was during the decade of geoeconomy that Japan acceded to the second rank among industrial powers, becoming the foremost investor and banker in the world economy. In counterpoint, the industrial decline of the U.S. superpower continued. An official report on the competitiveness of the U.S. high technology industry, published in March 1991—at the very moment when the highly symbolic Gulf War was coming to an end—concluded that in 15 of the 94 key technologies for the near future (in electronic chips and robotics in particular) American industry would no longer be present on the international scene in 1995. In 18 others, it was considered weak. It could only meet the challenge in 25 technologies, including data banks and artificial intelligence.[20]

Outside of finance and the automobile industry, the most spectacular penetration of Japanese firms has taken place in consumer electronics, where they control almost half the market. Starting from this locomotive sector, Sony and Matsushita have begun joining together equipment hardware and program software at the crucial moment when the fate of different high-definition television systems (HDTV) is being played out. In taking over two of the majors—Columbia and MCA-Universal—at the end of the decade, the two electronics giants have become owners of a quarter of Hollywood's studios. The early 1980s had already seen the Japanese animation industry displace U.S. producers of cartoons. Losing two ornaments of America's film industry led many Americans to feel that someone was stealing "a part of their national soul." So they expressed it in the polls that marked the rise of Nippophobia. This reaction was also fueled by a CIA report on Japan in the year 2000, the authors of which were not afraid to call on Westerners to be alert and mobilize against a country that was plotting to dominate the world and whose culture—"amoral, manipulatory and repressive"—would be "incompatible" with that of the "Judeo-Christian" West.[21]

One large problem for the Japanese cultural industry in its effort to establish itself on the world market in children's television was that it had to make use of the cultural heritages of others, for the most part Europeans.[22] In the same way, Japanese industry had borrowed the concepts of "productivity" and "quality" developed by Americans W. E. Deming and J. M. Duran in the period 1952–54 in order to develop their well-known model for managing the post-Taylorist corporation. And in conceiving its approach to the global consumer, Japan, possessing the world's largest advertising agency, had to make an alliance with other heavyweights in the triad. Japan thus occupies a leadership position in the production of the world economy, but a minor role in the "world culture" and in the cultural industry's representations of universality.

Finally, after a decade marked by economistic illusions and the fever of speculation, geopolitics, brought back onto the agenda by war, came to brutally remind Japan—but also Germany—that one can be the paragon of economic power in the twenty-first century and at the same time a political midget in the new planetary order. The old continent, for its part, could only note that the Europe of the Single Europe Act, already realized in the heads of those who conducted strategic studies on markets and transnational lifestyles, was far from constituting a solid and coherent political alternative to U.S. leadership, even if the United States was in industrial decline and dependent on Japanese chips to manufacture its Patriot missiles.

An indication of the difficulty of forecasting the outcome of the competition for world economic hegemony: in 1992, for the first time since they began to surge in the 1970s, the Japanese electronics firms suffered heavy losses in their turn. The same year, IBM, Toshiba, and Siemens announced that they were forming an alliance to perfect computer memories "16 times more powerful." This strategy, dubbed "double leapfrog" by the specialists, aims to jump over two generations of technology in order to take control of the following one. By allying to conceive the "chip of the twenty-first century," the three firms once again demonstrated the growing complexity of transnational cross-fertilization of capital and resources.

Chapter 11

Mediations and Hybridizations: The Revenge of the Cultures

Return to the Singular

Although the new philosophy of globalization inspired by geostrategies of the world economy came into vogue in the 1980s, there also emerged parallel visions of the world that were poles apart from such views.

As the "world system" deployed itself, connecting various societies with the products and networks destined to function at a universal level, other approaches to the transnationalization of culture, more concerned to restore its character as a process of multiple interactions, also came into being. What concerns us here are the responses of individual societies to the prospect of the reorganization of social relations brought about by new apparatuses of transnational communication that simultaneously destructure and restructure national and local space. These responses range from resistance to mimicry, from adaptation to reappropriation. In short, questions were now being asked about the processes of "resignification" by which the innumerable hookups to the networks that make up the fabric of globalization acquired a meaning for each community. Through this return to differences and to processes of differentiation, "international communication" is thus finally beginning to be conceived in terms of culture.

This is a return to cultures, territories, and particular spaces, as well as to concrete subjects and to intersubjective relations. Since the 1970s, critical research on the intercultural balance of forces had dealt above all with logics of deterritorialization and had favored the examination of the strategies of macrosubjects such as nation-states, major international bodies, or the new transnational economic groups—but also large-scale institutions representing the working class, such as parties and unions.

More recent critical approaches are attentive to the logics of reterritorial-ization or relocalization—that is to say, all the processes of mediation and negotiation that are played out between the singular and the univer-sal and between the plurality of cultures and the centrifugal forces of the world market, but also between different ways of conceiving the univer-sal. For despite the hegemonic concerns underlying mercantile concep-tions of cosmopolitanism, one of the major points of theoretical rupture consisted in exploding the essentialist conception of the "universal." Consequently, the very geography of social actors taken into account by analysis was called into question. New historical subjects began to inhab-it both theoretical references and reality itself. Other scientific disciplines were called upon and the monodisciplinary perspective was contested by cross-fertilizing viewpoints.

This sudden appearance of new ways of seeing not only the relation with the "international" but also, more generally, the relation with the "Other" took place in a context where theoretical ruptures had defini-tively lost any unambiguous meaning. The free circulation of skills and knowledge required by the new modes of social regulation introduced ambivalence as a major feature of the contemporary evolution of theory.

From Americanization to National Traditions

Two studies provide food for thought. The first is about the formation of the social group known in France as "*cadres,*" that is, business execu-tives. While published in 1982, it was mostly carried out in the 1970s: a time when business and its culture had not yet become the objects of a cult, and when numerous sociologists were concerned with the internal contradictions of corporations from a perspective not yet biased by the desire to master them for the needs of management. This study is the work of the sociologist Luc Boltanski of the Pierre Bourdieu research group. The second, undertaken in the second half of the 1980s, is the work of Philippe d'Iribarne, an engineer at the Ecole des Mines, a prod-uct of the elite Ecole Polytechnique, and director of the program "Tech-nology, Jobs, and Work" developed under the aegis of the CNRS, the French government research body. Its objective was to compare various "national traditions" of corporate management.

In a chapter entitled "The Fascination with America and the Import-ing of Management," Boltanski wrote:

One cannot understand the postwar transformations affecting the social representation of "*cadres*" if one is not aware of how much these changes owe to the importation of value systems, social tech-

nologies and standards of excellence of American origin which accompanied and sometimes preceded the realization of the Marshall
Plan, or, to be more precise, the political and social conflicts within
the bourgeoisie and petite bourgeoisie over the "Americanization
of French society." ... The introduction of human engineering and
of American-style management accompanied economic
changes. ... The enterprise of modernizing the economic apparatus is not just a technical matter. ... This modernization, partly inspired by the American economic authorities, who stipulated, as a
condition for the French to obtain funds, the formation of a group
of indigenous managers, economically competent and politically
reliable, presented itself explicitly as an endeavor to transform
French society as a whole.[1]

What Boltanski brings out is the process by which a class structure
was redistributed: how, putting multiple "social technologies" to work,
the group of *cadres*—a category that John Galbraith had been the first to
try to grasp under the term "technostructure"—became embodied in its
institutions and finally had its existence recognized as an objective, eternal, and natural fact.

Philippe d'Iribarne, for his part, endorses that nature of things. Inter-
rogating a sample group of managers in the United States, France, and
Holland, he tried to uncover the sources of efficiency proper to each culture. Writing from the point of view of one seeking to mobilize enthusi-
asm for the "economic war," he rebels against the discourses and global-
izing formulas taken from Japanese and American textbooks, and
counterposes to them the specifics of national cultures:

> What world are the advice-givers talking about when they people
> their books with an undifferentiated humanity—Japanese, Ameri-
> cans, or French, who, whether turners or accountants, are all melt-
> ed down into the same vague category? These advice-givers cer-
> tainly know that the passions that animate men of flesh and blood
> are often incomprehensible under foreign skies. How can they for-
> get that the traditions in which each people is rooted fashion what
> its members revere and despise, and that one cannot govern with-
> out adapting to the diversity of values and customs?[2]

Thus he points out, for example, that French managers obey a "logic of
honor" (hence the title of his book) that causes them to make an infinite
range of subtle distinctions between the noble and the base; that the
American responds to the lure of gain and the passion for honesty;
whereas the prudent Netherlander tries to bring different wills into agreement.

Through this reflection on the adaptation of management to individual conditions, a certain relativism is slowly introduced into the very heart of a science that sees itself as universal because it is "modern." This relativism is in tune with the search for organizational modes closer to the flexibility requirements of the network-corporation. Here is another passage from d'Iribarne:

> The research we have been able to do in the three "modern" societies of the United States, Holland, and France has not simply reminded us that modernity is not wholly triumphant and that traditions and particularisms are alive and well. It has profoundly transformed our perception of the relation between the modern and the traditional. . . . One can say that the contractual system is simultaneously modern and traditional, that it relies on ways of being, of living in society, that we are accustomed to attributing exclusively to either modern societies or to traditional ones. The relation it establishes between, on the one hand, structures and procedures, and on the other, traditions, is a relation of synergy, not of competition.[3]

This is not isolated research. From Africa to India via Mexico, numerous investigations in the areas of labor economics, labor sociology, and industrial economics began in the 1980s to focus on the types of organization best able to draw on local cultures in order to be more efficient.[4]

Simple Techniques

In deliberately drawing a parallel, almost at random, between two studies representing two different positions but also two moments in the way of defining the "international" and its relation to national cultures, we have deliberately tried to make apparent what the *problématique* of internationalization both won and lost in its passage from one perspective to another. Between the first and the second, the utopias that stimulated the desire for another kind of society had collapsed, and labor unions and working people's struggles had grown weaker, whereas management had redefined its ideology and strategy, regaining the means to think and act.[5] But this passage from one perspective to another also witnessed the rise to power of new professional strata, as the centrality of the working class and the workers' movement crumbled.

What separates these two outlooks is essentially the fact that the social technologies of mobilization, identification, and classification described by Boltanski present themselves more and more as "simple techniques." This occurs as the strategic place of these new strata, who play a mediating role in matters of power and in social conflicts, is legitimated within the corporation as well as in society.

The conflicts and divergences within the corporation over the very definition of "management," however, attest to the fact that there remains scarcely any neutral zone. More than ever, the corporation is a place of contradictory interests and of clashing value systems. This is shown clearly by the vicissitudes of introducing non-Taylorist forms of organization—in particular participant management—into the corporation. The logic of the corporation's "citizens" (in Norbert Alter's phrase), which tries to appropriate this mode of organization, collides with that of the "policymakers," who embody managerial logic and who see participation rather as a means of reaching an "efficient and happy consensus" and a "normalization of behaviors." These actors' logics, each of which represents a conception of the corporation, of the product, and of individual interest, conceive participation as a space of maneuver, of power relations, of struggles for influence. This leads Alter, a sociologist of organizations, to conclude that the "return of ideology" is one of the characteristics of the "modernist-participative" corporation, for the simple reason, in his words, that "once the notion of participation becomes the cornerstone of the functioning of this type of enterprise . . . , struggles become ideological wrestling, because culture and cultural influence give players muscle and the right to define the field as well as the nature of participation."[6] In the light of these hypotheses, one better understands the strategic positions acquired by communication and corporate culture as forms of social recognition in permanent tension between two ways of practicing participation: on the one hand the project developed by the management of formalizing a "responsible," "programmed," "rational," and "descending" participation, and on the other the project embodied by the group of corporate "citizens," in favor of "ascending" participation and "management of creative disorder."[7]

Félix Guattari attributes this reemergence of ideology to the necessity for the new agents of this development of "integrated world capitalism" to assure the construction of the subject and of subjectivity. The "structures that produce signs and subjectivity," he writes, have overtaken the structures that produce goods, causing all singularity to pass "under the domination of specialized equipment, professions, and frames of reference."[8]

The Management Horizon

To grasp the ultimate implications of this sanitizing of "specialized frames of reference" or "social technologies" one must assess what has changed in the ways of posing the issue of "mediators" or what might be called the "median class."

What has been eclipsed is quite simply any critical questioning of the role of carriers of knowledge and know-how in the redeployment of society and the economy. In the 1970s, this questioning was far from being the exclusive fiefdom of social critics. This is shown clearly by Michel Crozier, when, in his contribution to the Trilateral Commission's report on the factors in the "crisis of governability of liberal democracies," he expressed worry about the flagrant imbalance between the weight acquired by traditional intellectuals, seen as "troublemakers," and the other intellectuals who place their knowledge in the service of the functioning of society.

Crozier wrote in 1975: "A significant challenge comes from the intellectuals and related groups who assert their disgust with the corruption, materialism, and inefficiency of democracy and with the subservience of democratic government to 'monopoly capitalism.' " The attitude of these "value-oriented intellectuals," who devoted their energies to critiques and contributed to provoking a "breakdown of traditional means of social control, a delegitimation of political and other forms of authority," contrasted, by Crozier's own admission, with that of "the also increasing numbers of technocratic and policy-oriented intellectuals."[9]

At the root of this diagnosis of the "delegitimation of the traditional means of social control" was the idea—soon to gain currency—that the media had a great impact on the evolution of opinion concerning the Vietnam War, and also had influenced the memory of the challenge to that war in the West, in a context of generalized protest against the international system and the social order. Zbigniew Brzezinski, prefacing the report of Crozier and his colleagues, spoke of widespread disarray, and of pessimism regarding the evolution of Western societies: "In some respects, the mood of today is reminiscent of that of the early twenties, when the views of Oswald Spengler regarding 'The Decline of the West' were very popular."

To mitigate this crisis afflicting the very form of the democracies of North America, Europe, and Japan and making them "ungovernable," the 1975 report made the following recommendation:

> In due course, beginning with the Interstate Commerce Act and the Sherman Antitrust Act, measures had to be taken to regulate the new industrial centers of power and to define their relations to the rest of society. Something comparable appears to be now needed with respect to the media. Specifically, there is a need to ensure to the press its right to print what it wants without prior restraint except in most unusual circumstances. But there is also the need to

assure to the government the right and the ability to withhold information at the source.[10]

Since history often travels by detours, it would take more than fifteen
years for this prophetic recommendation to be realized, with the shack-
ling of the news media by the military authorities during the Gulf War of
January and February 1991—as I have already pointed out. To justify it,
the authorities of military censorship would invoke precisely the need to
not repeat the media excesses of the Vietnam War. Between the Vietnam
War and 1991, deregulation, actively consented to by the state, had
shown that it was not necessary, in order to regulate the "dysfunctions of
democracy,"[11] to resort on an everyday basis to the direct intervention of
state power; to find new "means of social control," it was sufficient to
shift society's center of gravity to an entrepreneurial logic.

The rise of the so-called postindustrial society only accelerated the upsetting of the relationship between traditional intellectuals and the others, causing the balance to tilt definitively to the side of the latter. In
1967, Daniel Bell, in his initial analyses of this type of society, had foreseen that it would be characterized as follows: "The whole structure of
social prestige is rooted in intellectual and scientific communities."[12] As
we have already stated, however, Bell was convinced that this ascension
of the new bearers of knowledge and know-how could only be accomplished under the protection of the state. The information society circum-
scribed by the market caused this predicted outcome to give way gradually to the logics of management and managers.[13]

As a last resort, all these ideological evolutions—since one must dare
to call things by their names—resulted in the undermining of the idea
that we were entering what the philosopher Gilles Deleuze called, after
William Burroughs, "control societies," and what Michel Foucault called
"disciplinary mechanism" societies, to differentiate them from the old
"discipline-blockade" societies. The crisis of the corporate regime as a
milieu of disciplinary confinement is the same as in all closed spaces (pris-
ons, schools, hospitals). The resolution of the crisis in one site often
accompanies its resolution in another, as if there, too, integration were
becoming the norm. The introduction of the corporation into all levels of
schooling is merely the most flagrant illustration of this. The forms of
these sociotechnical mechanisms of flexible control and the speed at
which they are instituted vary little from one institution to another, although the factory, contrary to what the hegemonic culture of leisure
might cause us to think, constitutes, at this stage at least, the model.

Gilles Deleuze, who is poles removed from enchantment with managerial communication and cultural discourses on the transformation of the corporation, writes:

> The family, school, army, factory, are no longer distinct analogous milieux which converge in a proprietor, state or private, but rather coded instances, deformable and transformable, of the same enterprise, which has only managers. . . . We are taught that business corporations have souls, which is indeed the most terrifying news in the world. Control is short-term and in rapid rotation, but also continuous and unlimited, whereas discipline was long-lasting, infinite and discontinuous. Man is no longer imprisoned, but indebted. It is true that capitalism has kept as constant the extreme misery of three quarters of the world, too poor for debt, too numerous for imprisonment; control will not only have to confront the dissipation of frontiers, but also the explosions of the slums and ghettos. . . . The rings of a snake are even more complicated than the holes of a mole-hill.[14]

Whether one likes it or not, the era of the information society and industry is also (when we look beyond the myopic gaze of its prophets) the production of mental states, the colonization of the mind. This requires us to think through in a different way the question of freedom and democracy. Political freedom cannot remain simply the right to exercise one's will. The increasingly fundamental problem is how that will can be formed. Unless we abandon the well-established belief that the fate of democracy resides completely in the media, we can scarcely hope to begin answering the question left up in the air by Deleuze, regarding the "gradual and diffuse institution of a new regime of domination" and the uncertainty concerning forms of resistance. This is true as well for intellectuals, who are more and more in the grip of managerial positivism, a new utilitarianism that stimulates the search for epistemological tools capable of circumscribing the potential zones of conflict and diminishing tensions through technical solutions.

An Alternative Modernity?

The theories of linear modernization expressed the Western vision of modernity. Its predictions were not realized. The political and economic forms inspired by modernization/development all failed.

Taking note of this patent failure, anthropologists started to investigate critically how political scientists, historians, and sociologists had posed the relation between transnational cultural flows and "national" cultures in the so-called Third World. Their central hypothesis was that the intensification of the circulation of cultural flows engendered by the process of transnationalization does not lead to a homogenization of the globe, but to a world more and more "hybridized." (Some preferred the term "creolized.")

Two scenarios seemed possible to these anthropologists regarding the long-term effects of transnational cultural flows. Either the transnational cultural mechanisms will continue to weigh indefinitely on the sensibilities of peoples on the periphery, subjected more and more to imported meanings and forms from which local cultures will hardly distinguish themselves, to the point of blending with them. Or else, with time, the imported forms will be tempered and recycled by local cultures. In reality, these two scenarios are woven together.[15] The Indian anthropologist Arjun Appadurai writes:

> The globalization of culture is not the same as its homogenization, but globalization involves the use of a variety of instruments of homogenization (armaments, advertising techniques, language hegemonies, and clothing styles) which are absorbed into local political and cultural economies, only to be repatriated as heterogeneous dialogues of national sovereignty, free enterprise and fundamentalism in which the state plays an increasingly delicate role: too much openness to global flows, and the nation-state is threatened by revolt—the China syndrome; too little, and the state exits the international stage, as Burma, Albania and North Korea in various ways have done. In general, the state has become the arbitrator of this *repatriation of difference* (in the form of goods, signs, slogans and styles). But this repatriation or export of the designs and commodities of difference continuously exacerbates the internal politics of majoritarianism and homogenization which is most frequently played out in debates over heritage.[16]

Appadurai even dared to speak of "alternative modernity." This idea is borne out by studies of advertising and the building of "consumer communities" in India where, contrary to what had happened in the West where the ideology of nationalism had preceded the arrival of advertising techniques, the development of advertising took place contemporaneously and synergetically with it. These studies noted the rapid emergence of middle classes with a significant disposable income and cosmopolitan tastes, and the explosion of efforts on the part of businessmen to diminish the gap between signs and dream and between products and markets.

In a similar vein of research, Brazilian anthropologists undertook to retrace the history of the "modern tradition" in their country, in the words of Renato Ortiz, taking as their thread the genesis of the cultural industry and the national market for cultural goods.[17] It is an admixture of the modern and the traditional, as witnessed by a remarkable alloying of mass culture and popular cultures in the products of Brazil's highly competitive television industry, which succeeds in combining postmodernity and signs of the preindustrial.[18] Brazil even outperforms countries such as France in the world market for programs. This leads Ortiz to conclude his study with these words:

> The debate on the national takes on a different significance. Until now, it was limited to the internal frontiers of the Brazilian nation. . . . Today, it is transformed into an ideology to justify action by corporate leaders on the world market. This is no doubt the reason why there has been no major difference between the sales discourse for the *telenovela* and the arguments of arms dealers on the export market (Brazil is the fifth world producer), since the two are seen exclusively as national products. I would say, then, that this marks a new stage in Brazilian society, making it impossible to return to the old opposition between colonizer and colonized with which we are accustomed to operate.[19]

To put this so-called alternative modernization into perspective, and to avoid being deceived by a new myth, we have still to complete this vision of modernity, based on the progress of classes and social categories integrated into its system of rewards, with other logics—those of segregation, which within the same social reality have not ceased to grow. To characterize it, Spanish sociologist Manuel Castells has even employed the notion of "new dependence." The new model of world development provokes a progressive detachment of segments of economies, cultures, societies, countries, and social groups, all of which cease having a functional and economic interest in the system as a whole, being too poor to constitute markets and too culturally backward to serve as a work force in a productive system founded on information. All societies in search of an "alternative modernity" are also societies of uncontrolled modernity—uncontrolled for those segments that remain outside the global economy functioning as a unit in daily real time. "The basic process experienced by what we call the Third World," writes Manuel Castells,

> is its disintegration as a relatively homogeneous entity. South Korea or Singapore is closer to Europe in terms of economic and social development than to the Philippines or Indonesia. Even more important is the fact that São Paulo is further away socially from Recife than from Madrid. And that in the state of São Paulo itself, the Avenida Paulista and the working-class city of Osasco belong to different socioeconomic constellations, not only in terms of social inequality, but also in terms of cultural dynamics and segmentation. . . . Therefore, the story begins in the context of the seg-

regation of a large portion of the population of the planet, not in the dangerously simplifed terms of North vs. South, but in a more complex and insidious way.[20]

This logic of social segmentation is taken for granted and accepted by the heads of global advertising and marketing firms, who do not hesitate to draw strategic lessons from it.[21]

The sectors of society thus rejected by the global network of discriminatory "interdependence" supply recruits to the "new front of planetary disorder," in the phrase of the geopolitical specialist Michel Foucher,[22] with its many expressions, such as drug trafficking and money-laundering; collective explosions of looting, which target the sanctuaries of consumption; the irrational appeals to cultural, ethnic, or religious identities; the rise of fundamentalisms that, paradoxically, have incorporated into their strategy both the most modern means of communication and the innumerable suicidal weapons of the excluded, not to mention the rise in would-be political saviors, packaged with the aid of the audiovisual media, who fill the void left by the crisis of political representation.

But above all, there is a new everyday disorder that has turned the culture of violence into a normal dimension of life. A Peruvian theater critic wrote in 1990 that "the crisis has ended up inhabiting the most intimate sphere of our daily life. The pauperization and the semi-proletarianization of the middle class, as well as the fall in the consumption level of the popular strata, have produced a series of spontaneous responses to the crisis, giving rise to a new urban, social, and cultural ecology, in which we begin to see our own reflection."[23]

These everyday experiences are precisely what transnational communications apparatuses rarely convey, since they are more accustomed to covering police operations against drug trafficking than to expressing how Third World people might continue to maintain their dignity, even though every threshold of violence has been exceeded.

These media representations of the other as agent of the "fronts of disorder" lodged themselves in a social reality in which the way of experiencing the Third World has changed completely. The heroic idea of the faraway Third World of the 1960s has given way to a Third World represented as the experience of minorities on the home territory, product of the great diasporas of the labor market. It is an experience bristling with sensitive zones, always ready to explode into crises, into cultural conflict, against the background of rising political and religious fundamentalisms and fanaticisms. The situation has become all the more complex in that the extreme right has turned inside out the right to difference—what Pierre-André Taguieff calls "differentialism," a touchstone of anti-racist thought—and now uses it to preach exclusion of the "foreigner" and the revival of the idea of the nation as patrimony and fatherland.[24]

Rethinking Popular Genres

The anthropologists' new formulations about the possibility of an "alternative modernity" suggest the aptitude of societies and their various components to deviate, corrupt, and pervert the instruments by which this difference was relegated to the margins.

This line of research, which expresses a movement toward the reappropriation of particular histories, also inspires studies that in the 1980s began to examine the unequal exchange between mass culture and the experience of popular cultures. This is especially true of research on television genres of national or regional origin, and particularly genres belonging to the great tradition of melodrama as it appears in Egyptian serials, Latin-American *telenovelas*, or the cinema of India, the most prolific producer in the world.[25] What this sort of study is trying to understand is the capacity of this type of narrative to create veritable states of catharsis on the scale of an entire country, if not a whole continent, and to mobilize its effects. Beyond case studies, a whole field of investigation has opened up on the formation of national identity and national-popular cultures, a question already taken up by Antonio Gramsci in the 1920s; on the confrontation between these cultures and transnational networks; and finally, on the role of intellectuals in all these processes of acculturation.

Another dynamic that contributed in many Third World countries to the renewal of theoretical questions was the rise of communication networks and popular or "participatory" education. This kind of communication, which employs various media from video to radio and the traditional printed forms, goes hand in hand with the search for types of self-organization by which new social actors try to assume the management of their own affairs, in a context where the state has ceased to provide welfare, if it ever did. Of course, these microexperiences do not always avoid the trap of "basism," a constitutive element of the history of quests for "alternative communication" everywhere. ("Base" is used here in the sense of "grassroots"; "basism" is thus an exaggerated concern with the preferences and desires of participants.)

In any case, the point to be emphasized about the rise of the networking conception of social organization, as implemented by diverse nongovernmental organizations, is that it has begun to stimulate new forms of international exchange between North and South and between South and South, originating from civil society. This has the merit of laying the basis for reflection on an "international third space" (as in Third Estate) that, if we were to daydream, might find a place between intermarket logics and interstate logics that mediate respectively the pragmatism of the merchant and the *Realpolitik* of the prince fettered by the reason of state.[26] This reflection is all the more crucial in that to the search for a redefinition of North-South relations has now been added a search to redefine those between East and West. There is a need to construct ties other than those determined by the expansionary logic of a world market already too accustomed to reducing freedom to the freedom of commercial expression and citizens' rights to consumer sovereignty.

Solitary Pleasures

As we have already noted, in the redeployment of free enterprise the consumer is the keystone. He or she is at once, as "coproducer," one of the links in the production process and, as representative of the people-as-market (*peuple-marché*), the key to the process of legitimation of the neoliberal conception of society. For it is not a matter of just *any* consumers, but rather of consumers who are sovereign in their choices in a free market. Neoliberalism, in its struggle against all forms of control (except of course its own, those of free enterprise), whether they emanate from the state or from organized civil society, reveals itself as a form of neopopulism as well. Thus it experiences the constant need to reaffirm the representativity of consumers in their role as market shares. It speaks in their names. Hostage and alibi, this consumer has, indeed, the starring role on the stage of the democratic marketplace; he or she is a "citizen" of it. The discourse built around the consumer, a consumer free of all attachments and determinations other than his or her own will, claims such authority that it often becomes a totalizing discourse, one leaving no place for other issues than those related to consumption. Consumption is assumed to contain within itself its own explanation and *raison d' être*.

This logic, which seeks to rehabilitate the consumer and represents a new situation in societies subject to the laws of the market, does not facilitate the critical apprehension of the different and contradictory theoretical movements that developed since the early 1980s around the status of the consumer as receiver or user of the media and of communicating machines. In an age when the theme of reception is quite widespread, so frequent are the efforts to make us believe that the return to the consumer is necessarily interesting in itself and constitutes a fundamental break with the past, that one often forgets to question the reasons for the evolution of these approaches and the origin of their diversity. To justify the celebration of this return to the consumer and user, however, it does not suffice to cite a whole series of studies illustrating this phenomenon. For those who do not wish to dance exclusively to the "reception" tune, it is necessary to clear up a few ambiguities, and this requires us to return to the past.

Who can deny that the problem of reception upset the old-style functionalist research? As Wilbur Schramm admitted in 1983, shortly before his death: "It is illuminating to think of communication as a *relationship built around the exchange of information*. The process of exchange is more likely to resemble a biological than a physical one, to make a difference in both parties, to change the relationship rather than changing one participant."[27]

But it is to Elihu Katz that we should turn for an example of research that claimed to be innovative in the field of reception studies from a functionalist perspective. In the course of the 1980s, Katz and his collaborators performed a series of investigations in an effort to discern the different ways in which receivers of diverse ethnic origins, such as Palestinians living in Israel, Moroccan Jews, and Americans of Los Angeles, decoded the television serial *Dallas*, the product par excellence of a culture supposedly universal in range. Their main observation was that the readings were very diverse, and that each reading was a function of the nature of the viewer's implication in the serial, which was related, in turn, to the manner in which each viewer's culture constructed the role of the viewer. The study in itself combines empiricist attention to detail and a certain myopia caused by its theoretical poverty.[28] From this point of view, the judicious criticism addressed to Katz by James W. Carey and James Halloran in the late 1970s, when his expertise was sought by the BBC, had said everything there was to say regarding the merits and failings of this type of approach.

For this research adopts a conceptual framework that is not so new. It is the culmination of a long evolution in which two things are at stake, since it operates on two fronts. The first is the one opened up in the 1950s by the theory of the *two-step flow*, and is directed against the Lasswellian theory of the media ("Who says what to whom on which channel with what effect?"). To the question: "What effects do the media produce on society, groups, and people?," Katz counterposed another interrogative chain: "What do people, groups, and society do with the media?" This was the question that became more and more crucial to him beginning in the 1970s, to the point where it fueled a whole current of studies on the satisfactions of media users, which would be known as "uses and gratifications." These studies developed the notion of "negotiated reading," that is, a reading or reception in the course of which meaning and the effects it produces are born of the interaction between the program and the roles assumed by different types of viewers or readers. This new line of research would have broad repercussions not only in the United States but also and most notably in the United Kingdom.[29]

The second front aimed to refute the very idea of power as it had been developed by various critical traditions. This aim became increasingly important, as one could not avoid coming to terms with these traditions in posing the problem of international communication, and also because Lasswellian visions of the media were in decline. We need not insist on the disarray experienced by empiricist sociology in the face of the "dysfunctions" occasioned by the debates on the new order and national communication policies and the absence of a discourse or analytic framework other than those closely complicit with the neoliberal philosophy of *free flow of information* and the free consumer in the free market—the free fox in the free chicken house.

Yet one may address to Katz's reception studies exactly the same criticisms that Jean-Marie Piemme formulated in 1978, in a book that remains a classic, about the uses and gratifications theory: "With Lasswell as well as Katz," he wrote,

> one finds juxtaposed two elements (the media on one side, people, groups, and society on the other) which are assumed to be autonomous, then one asks about their relation. This clearly means there was no initial effort to locate the media within the circumstances of the social formation. They are resented outside any structure and appear to engender on people/groups/society an effect *sui generis,* whose structural determinations and contradictions are ignored, as if they had nothing to do with the power relations that give the social formation its particular configuration. This theory seems oblivious to the fact that the media take part in social contradictions and that their effects are interventions that may either comfort or alter the existing balance of forces.[30]

All this has been known for a long time. The novelty resides in the role demanded of these studies—a role they intend to play[31]—in the current context of confusion, characterized by the lack of questioning of their epistemological status. Why continue to speak of power relations between audiovisual cultures and economies when the way people decode this supposedly "ecumenical" serial (but not *timeless*, since the producers of *Dallas* decided to suspend it in 1991!) proves that they have a formidable power, that of conferring meaning: here is an argument, drawn from this type of study, that serves to shunt aside all the issues raised in the conflictual history of communication, its theories and their uses, ever since the first concept of communication was forged.

Such issues are also passed over in silence by the ethnographic variety of research on audiences, which flourished in the 1980s, particularly in major industrial countries where old conceptions of vertical and monolithic power had been abandoned. It, too, isolated the problem of reception while exaggerating the power of receivers and overestimating the value of the direct, negotiated encounter between "offer" and "demand." While claiming not to belong explicitly to the functionalist tradition, in fact this type of ethnographic research revalidated its premises and endorsed, like other currents, the idea of the absolute freedom of the consumer in the "choice of meaning."[32]

Following on the determinist conception of the abstract consumer dear to the structuralist research of the 1960s and 1970s—in which the consumer, lacking a voice, was subject to the imperative of a structure, and the broadcaster was no less abstract—this new version of empiricism brings us receivers so concrete that one forgets which society they live in, thus concealing both the tree and the forest. And this is happening at a time when the process of Taylorization of consumption is making palpable the need for information about a receiver who is both the point of departure and the end point for strategies of globalization and the infinite segmentation of targets and markets. It was as if, in this contemporary hybrid reality, ethnography retained only one strategy, that of personalization: the contingent, the singular against the global. What is clearly wrong with this approach is not so much its attention to the minute detail of procedures, a quality one can only praise because it undermines the old determinisms of all stripes, as its willful or forced extrapolation to the world-space in order to deny the latter's inegalitarian logics.

This ethnographic vision is a good partner for the postmodernist conception of audiences, in which the television viewer, in insouciant nonchalance, operates shrewdly in media spaces. In this universe of nonsense where everything is equivalent to everything else, the new relativism completes its cycle.[33] It is careful not to ask why certain problems are well studied and others are less so, or even totally ignored.

Tactics

The new empiricisms of consumption fortunately do not exhaust the theoretical and practical upheavals that have occurred in this field. There are other responses to this crisis of the macroscopic conception of power. Competing traditions of research have formulated differently the problems of the reception of television programs and the construction of their meanings by the public(s).

Here, there is definitely a break with preceding decades. What is new is that the problem of reception and use has become impossible to evade. Well before it was widely recognized as important, this concern for reception and uses had its pioneers. For example, there was the British author Richard Hoggart, whose first book, published in 1957, was significantly entitled *The Uses of Literacy*. It was one of the first studies of the evolution of the culture of the popular classes under the influence of modern mass culture.[34] Hoggart himself belongs to the antipositivist tradition of the study of cultural forms and relations initiated in the 1920s by F. R. Leavis.

Another venerable tradition of literary research raised questions about the role of receivers long before the sorcerer's apprentices of communication. It defined literary work metaphorically as "encounter" (R. Ingarden), "dialogue" (M. Bakhtin), "convergence" (H. R. Jauss), or "interaction" (W. Iser) between the text and the reader.[35] Jean-Paul Sartre, whose role as precursor in the reorientation of literary studies is undeniable, made the relation between the reader and the text the essential starting point for answering the question "What Is Literature?" (1947).

The institution of the theme of reception and uses, or of the receiver and users, as the norm in communication research, took place under the effect of very different logics. We have already alluded to the industrial ones. Social logics, insofar as it is possible to separate them from the latter, refer to the new conditions under which democratic life is organized. For those who wish not to reduce the problem to an equation of supply and demand, the theme of the active role of the receiver and user is indissociable from questions raised by citizens organized in civil society about the possibilities of exercising a real democratic control over the new flows and new networks. The notion of use concerns not only what one can do with television programs, but the utilization of all the technological tools of the new mode of communication. We are beyond the stage of posing the problem in the simple and unequivocal terms of state control over such tools.

It would be illusory to look for a single body of critical knowledge that systematizes this return to the user. At best one can sketch some traits that identify the origin of these new hypotheses and distinguish them clearly from the new empiricist currents.

The point of departure of the new critical theory of social uses is first of all a position confronting the idea of order and control with the reality. Its corollary is the idea that this order—that of the state and the market—and its multiple networks can be appropriated and diverted by its users, and that there is no passive consumption. The new critical thought on uses and users inevitably returns, then, to a conception of power and counterpower. If society is made up of apparatuses that produce control and constraint, adhesion and conformity, it is also made up of discreet ruses, inexhaustible tactics, subtle reappropriations and evasions, makeshift constructions, poachings, and unforeseen uses that preserve, even in submission, the inalienable freedom of the ordinary person, the "man without qualities," the "common hero," target of all efforts at domestication.

Michel de Certeau, historian, linguist, psychoanalyst, and ethnologist, made it his project in the late 1970s to advance some hypotheses about what he called the "invention of everyday life." In counterpoint to the strategies of Foucault, who analyzed society's networks of surveillance in *Discipline and Punish*, de Certeau speaks of "networks of antidiscipline," revealed by everyday practices or "ways of doing": those tactics or "operations" of users that constitute the active process by which, out of products and norms, they manufacture their own styles. Consumption becomes the art of using products. Statistical analyses of the time spent in front of the TV set, or the number of books sold, and content analyses of broadcasts, tell us little about the receiver, since there is no equivalence between the product disseminated and the product consumed. One must thus study the everyday practices of users in a logic of production or appropriation, no longer just in a logic of reproduction.[36]

Obviously we should not look to a historian for a methodological answer, in the American sense of the term, to the question of how to study daily practices. Intuitive and poetic, de Certeau's thought does not deliver tools for grasping these indomitable practices and thus it does not open the door to a control and surveillance of users by revealing predictable patterns of behavior. This is doubtless one of his virtues, at a time when the exhaustive study of audiences aims to measure the user's smallest acts and gestures in order to chart the inputs and outputs of cultural production. De Certeau's importance is elsewhere: it lies in providing a counterweight to those analyses that privilege invariant factors and social determinisms, and in reminding us that a common error is to analyze the effects of power by beginning with power itself, its acts and its perspectives, rather than with those who are subjected to it.

The trap here is clearly that of underestimating the importance of the great industrial and financial strategies, as well as the geopolitical stakes of industrial production of culture and communication. But Michel de Certeau does not see this as his concern. He leaves that to Foucault, whom he criticizes for having constructed an overly coherent system, while nonetheless subscribing to the new perspectives opened up by Foucault's examination of contemporary "technology of observation and discipline," the organizer of spaces and languages. And since in de Certeau's work, and contrary to the empiricist approach, the examination of practices does not imply a return to social atomism—his project being not psychological, but sociological, involving the study of modes of action and not the individuals who carry them out—there are no grounds for holding de Certeau responsible for a lack of analysis of the macrosocial outer lining of network society. For the person who goes to the trouble of seeking further, his decisive contribution consists not only in informing studies of "consumption practices" but also in obliging us to adopt a different outlook on the formation of apparatuses of mass cultural production.

"Fools imagine," wrote Marcel Proust, "that the large dimensions of social phenomena are a fine occasion to penetrate further into the human soul; they should understand, on the contrary, that it is by plunging into the depths of individuality that they have a chance of understanding these phenomena."[37]

The model of the plural should be sought in the singular: this claim was also made by Gabriel Tarde, for whom the history of societies repeats, on a larger scale, the history of individuals—against a Durkheim who postulated that the collective is not reducible to the individual. Nearly a century has passed, and this tension still exists. But the conviction that it is difficult to understand one without the other has at least cut tracks in the analysis of the supranational social relation, even if it has not opened broad avenues.

Conclusion:
The Enigma

Every attempt to retrace the history of international communication runs up against three major stumbling blocks, which also risk interfering with the reading of works on this subject.

First, there is the polysemy of the word "communication," torn as it is between the domains of leisure and work, between the spectacular and the ordinary, between culturalist and technicist visions, or tossed about between a meaning confined to the area of media activity and a totalizing meaning that elevates it into one of the basic organizing principles of modern society. Despite the centripetal force of the media representation of the phenomenon, it is tending more and more to the latter. In proposing to analyze the how and why of the different contents and uses assigned to the concept of "communication," this book shows what concrete issues were at stake in the various definitions in the course of its history, and why they continue to be so. This history of the concept leads us to conclude that only an analysis placed under the sign of culture can today account for those stakes. Culture is understood here as the collective memory that makes communication possible between members of a historically situated community, creating among its members a community of meaning (the *expressive function*), allowing them to adapt to a natural environment (the *economic function*), and finally, giving them the ability to construct rational argument about the values implicit in the prevailing form of social relations (the *rhetorical function*, that of legitimation/delegitimation). These are the three dimensions of culture that the philosopher Jürgen Habermas brings together under the trilogy *language, labor, power*.

Next, there is the danger of allowing oneself to be enclosed within the "international," just as some, at the other end of the spectrum, risk becoming immured in the ghetto of the "local." In succumbing to this danger, one risks subscribing to a determinist conception in which the international is converted into the *imperative*—just as, at the opposite pole, the exclusive withdrawal into the local perimeter is the shortest way to relativism. There is overestimation of the international dimension on one side, underestimation on the other. All these levels of reality, however—international, local, regional, and national—are meaningless unless they are articulated with each other, unless one points out their interactions, and unless one refuses to set up false dilemmas and polarities but instead tries to seek out the connections, mediations, and negotiations operating among these dimensions, without at the same time neglecting the very real existence of power relations among them. This mode of observation and analysis is far from having been the norm in the history of theories of "international communication." This explains why and how certain conceptions of the international have offered local and national powers a convenient set of justifications for evading their own responsibilities, by shifting them to some faraway supranational determination. Conversely, the dismissal of the international dimension and flight into narrowly defined identities have often been complicit with the most extreme nationalist ideologies of exclusion of the foreigner. This book, in examining the dynamics that have, in every period, given rise to novel relationships between the "foreign" and a given society, may also be read as a history of the concept of the "international."

The eras of the *Pax Britannica*, the *Pax Americana*, or the *Pax Sovietica*—the era of states inclined to prophetic visions of their own grandeur and the unshakeable affirmation of their superiority—gave rise to the tendency to look at the world from the point from which power radiated outwards. The East-West confrontation has left its imprint in the form of a bipolar division of the planet that fueled the imaginary with a metaphysical contest between the forces of good and evil—at least until the day when the bloc conception crumbled along with regimes thought eternal and omnipotent. And yet the Manichean vision of the planet has not vanished from mentalities. The Cold War had scarcely been buried when a regional war broke out, and this religious conception of grand international oppositions made a spectacular resurgence. The havoc it has wrought is visible even among the most enlightened intellectuals.

But the idea of mediation has begun to defeat the conceptual arsenal of the era of glaciation. Memory is fading of the time when one interpreted the procedures of internationalization as if they operated like a steamroller, a deus ex machina sweeping away everything in its path and succeeding in covering over with its totality all individual and collective life. If we now recognize that each society has the mode of access to internationalization that it deserves, it is because in the meanwhile, the idea has been accepted that encounters and exchanges between singular experiences and the supranational, while often unequal, are also processes, that is, social constructions. The crisis of the grand paradigms that oriented social change and guided the social sciences is the crisis of centrality. And as such it signals the decline of the imperial modes of administering and controlling. The multipolar world that weaves its networks across the planet has made more complex the forms of subordination of some societies to others, some cultures to others, some ways of life to others.

The archaeology of concepts doubled with that of facts undertaken in this book finds its justification in the relative void in the history of communications. Forgetfulness of history is, in fact, one of the recurring traits in thought about communication. This forgetfulness explains, for example, why for so many years the debate over the media was polarized between the "apocalyptic intellectuals," who denounced the media as bearers of "the end of culture," and the "integrated intellectuals," who celebrated the same media for their "modernizing" virtues. Both sides, unwittingly, gave new validity to the myth of the media's omnipotence, until the day when the ever more massive presence of the media in society provoked a crisis in the minds of many intellectuals who had taken refuge in the lofty solitude of creation, forcing them to question the elitism of certain positions they held, often without bothering to consider the viewpoint of the mass of consumers and users. This amnesia was equally at the origin of economistic illusions and excesses that attributed to networks and technologies of electronic communication miraculous powers to restructure economies and rebuild consensus—the democratic transparence of the informational agora discovered at last.

It is only recently that one has begun to sense, in many parts of the world, the need for a critical history in this field. For the moment, though, historical research is manifested mainly in the form of a return to national histories, while the international is still left by the wayside. This new consciousness of the necessity of finding historical roots converges with an awareness of the need to establish an epistemological reflection in this scientific field that is still young and uncharted. The field is also disturbed by the noisy clamor of the ephemeral, one piece of information eclipsing another, a situation that makes the intellectual feel guilty for refusing to be converted into a media pundit appearing live to discuss every event. The infatuation in the 1980s with the notion of communication as progress and the escalating predictions of technological innovation led many to believe that history could be done without. Geopolitics quickly took its revenge on a speculative geoeconomics that, disconnected from the real world, thought it had become autonomous. Consequently, the enchanted discourse on communication has grown old and wrinkled, although it would be the height of temerity to celebrate its demise. In any case, it is up to us to decide whether this questioning will go further. "Communication" has indeed become too important for the future of relations among cultures to go on deferring its reappropriation by ordinary citizens.

A third and final danger: Can the reader or the author of an international history escape the ethnocentrisms lying in wait on all sides? This comes down to asking: From what socially and historically given territory can one speak of these phenomena at the dawn of the third millennium? It is difficult to avoid this question at a time when some celebrate the fifth centenary of the "discovery" of America by Christopher Columbus, or the "encounter of two worlds," while others, with respect to the same event, commemorate the fifth centenary of the "birth of a historic system of injustice" and pay homage to the memory of victims of "one of the largest pillages and genocides in human history."[1]

This is a question that in the 1950s already preoccupied the European intellectual Elias Canetti, who was born in 1905 in Bulgaria of Sephardic Jewish parents, studied in Zurich, Frankfurt, and Vienna, and became a refugee in England in 1938. In his key book, *Crowds and Power*, in which he analyzes "national ideologies" and their influence on the theoretical apprehension of the foreigner, he wrote: "One must stand apart, a devotee of none, but profoundly and honestly interested in all of them. One should allow each to unfold in one's mind as though one were condemned actually to belong to it for a good part of a lifetime. But one must never surrender entirely to one at the cost of all the others."[2]

The formidable growth in the means of communication and the instances of contact between cultures that has come about since the 1950s has not changed the problem in the slightest. It has merely extended it to the frontiers of the world-system by multiplying the number of actors who try to resolve what risks remaining an enigma for a long time to come. With the rise of the multiethnic and multicultural challenge faced by societies throughout the world and with the ensuing threat of withdrawal into narrow identities, we are reminded, in the first years of the 1990s, that history has not granted the idea of "nation" an unequivocal meaning, and that different "nationalisms" have very different political and cultural meanings.

In these times when, more than ever, we should protect ourselves against an ethnocentric quest for collective identity, there is no better way to do so than with the words of the Argentinean Jorge Luis Borges:

To set up a worldwide organization is no trifling enterprise. . . .
Twirl, who had a farseeing mind, remarked that the Congress involved a problem of a philosophical nature. Planning an assembly
to represent all men was like fixing the exact number of Platonic
types—a puzzle that had taxed the imagination of thinkers for centuries. Twirl suggested that, without going farther afield, Don Ale-
jandro Glencoe might represent not only cattlemen but also
Uruguayans, and also humanity's great forerunners and also men
with red beards, and also those who are seated in armchairs. Nora
Erfjord was Norwegian. Would she represent secretaries, Norwegian womanhood, or—more obviously—all beautiful women?
Would a single engineer be enough to represent all engineers—including those of New Zealand?[3]

Appendix:
Selected Chronology

1785　First issue of the *Times* newspaper in London.

1788　In the *Times* of January 1, the foreign news from Rotterdam and Paris is dated December 25, 1787; the news from Frankfurt, December 14; and from Warsaw, December 5.

1790　First census in the United States.

1792　The French commission of weights and measures begins to work on the definition of the metric system.

1793　Inauguration of the optical telegraph in France (Chappe).

1798　Landing in Alexandria of General Bonaparte's expeditionary force, accompanied by an auxiliary corps of experts and printers. Publication of *An Essay on the Principle of Population* by T. R. Malthus.

1801　First census in England and in France.

1814　First application of the principle of steam-operated printing press (Koenig and Bauer) for the London *Times*.

1821　Publication of *Le système industriel* (The industrial system) by Saint-Simon.

1822　In France, "la fille ainée de l'Eglise" ("oldest daughter of the Church"), Lyon, becomes the center for the Labor of Propagation of the Faith in infidel countries; its *Annales* are translated into several languages; the beginnings of the modern Catholic missionary press.

1825　First important railway line (Great Britain).

1827　Photography (Niepce).

1829   Steam locomotive "The Rocket" (George Stephenson).

1830   Comte begins to develop positive philosophy course.

1832   Publication of *Le système de la Méditerranée* (The Mediterranean system) by the Saint-Simonian Michel Chevalier.

1833   First issue of the *New York Sun.*

1835   Creation of the Havas news agency.

1837   Electric telegraph (Morse and Cooke-Wheatstone).

1838   First daguerreotype (Daguerre).

1840   Invention of the adhesive postage stamp and reform of the postal service in England (Rowland-Hill). Creation of the first modern U.S. advertising agency.

1842   Decision to create major railway lines in France.

1845   First issue of *Scientific American.*

1846   Two-cylinder rotary press (Hoe).

1847   Postal service reform in the United States.

1848   Postal service reform in France. Publication in German of the *Communist Manifesto* by Marx and Engels. Creation of the American news agency AP (Associated Press).

1849   Creation of the German news agency Wolff.

1851   France-England underwater cable. Creation of the Reuter news agency. Crystal Palace Exhibition in London.

1852   Publication of *Social Statics* by Herbert Spencer.

1855   Cable from Varna to Balaklava (Crimean War). Universal Exposition in Paris. Abolition of the stamp duty in England.

1856   British decree regulating the relations between the press and military (Crimean War).

1859   Publication of *On the Origin of Species* by Darwin.

1861   Abolition of the paper tax in England.

1862   Exhibition in London.

1865   Foundation of the International Telegraph Union. Creation of the U.S. advertising agency J. Walter Thompson.

1866   First operational transatlantic cable. Refinement of the rotary press (Marioni and Bullock). First typewriter.

1867   Publication in Hamburg of the first volume of Kapital by Karl Marx. Paris Exposition.

1869   Creation of the U.S. news agency APA, later UP and then UPI (United Press International). Opening of the Suez Canal. First transcontinental railway in the United States.

1870   Sharing of the world market among the news agencies of the cartel (Havas/Reuter/Wolff).

1871   Underwater cables in China and Japanese seas.

1873   Refinement of the Remington typewriting system. Vienna Exposition.

1874   Cable network in the South Atlantic. Dissolution of the First International. Presentation of the report on "public services" by C. De Paepe. Herbert Spencer begins the publication of *The Principles of Sociology* (1874–96).

1875   Beginning of the "Age of Empire" (1875–1914). Elisée Reclus begins the publication of *Nouvelle géographie universelle* (New universal geography) (1875–94). Creation of the International Bureau of Weights and Measures. Founding of the Universal Postal Union.

1876   Invention of the telephone (Bell). Universal Exposition in Philadelphia. Publication of *L'uomo delinquente* (Criminal man) by C. Lombroso.

1877   Commercial exploitation of the telephone by Bell Telephone Company.

1878   Invention of the phonograph and plan for electrification of New York (Edison). Gelatin-bromide photographic plate (Eastman). Printing telegraph (Baudot). First telephone lines (United States). F. W. Taylor, the father of Taylorism, hired by the Midwale Steel Company. Universal Exposition in Paris.

1880   Invention of the perforated card (Hollerith). Publication of *Le droit à la paresse* (The right to be lazy) by Paul Lafargue.

1881   French law on freedom of the press. International Exposition of Electricity in Paris.

1882   International Conference on railways. Establishment of the first subsidiary of the future International Western Electric factory network in Anvers, Belgium (Bell Manufacturing Company).

1883   Founding of the Alliance Française.

1884   Improvement of the camera lens (anastigmatic). Adoption of world time (Greenwich Meridian).

1885   Founding of Gaumont. First international congress of criminal anthropology. Creation of the International Statistics Institute. Apogee of the *feuilleton* genre—inaugurated in 1836—in the large-circulation papers in France.

1886   Invention of the linotype (Mergenthaler).

1887   Interstate Commerce Act (United States). Publication of *Gemeinschaft und Gesellschaft* (Community and society) by Ferdinand Tönnies.

1888   First Kodak camera (Eastman). Founding of the *Financial Times* (London).

1889   Founding of the *Wall Street Journal*. Creation of the Second International under social-democratic control. Universal Exposition in Paris.

1890   The French popular daily *Le Petit Journal* reaches a circulation of a million copies. Use of perforated-card machine in the U.S. census. Publication of *The Influence of Sea Power upon History* by A. T. Mahan. Publication of *La photographie judiciaire* by A. Bertillon. Publication of *Les lois de l'imitation* by G. Tarde.

1891   Publication of *Method of Indexing Finger Marks* by Sir Francis Galton.

1893   Radioelectric antenna (Popov). World's Columbian Exposition in Chicago.

1894   First comics in the Hearst and Pulitzer newspapers. Publication of *The Theory of Transportation* by C. H. Cooley.

1895   Publication of the *Psychologie des foules* (Psychology of crowds) by G. Le Bon. First cinematographic exhibitions in Paris and Berlin.

1896   First steps in radiocommunications (Marconi). Color printing for comics (C. Saalburgh). Founding of Hollerith Tabulating Machines. First cinematographic exhibitions in London, Brussels, New York, and the major capitals of Latin America.

1897   Publication of *Politische Geographie* by F. Ratzel.

1898   The Dreyfus Affair in France, Franco-British crisis in Fashoda, and the Spanish-American War in Cuba: three events widely covered by the media. Founding of the British firm Marconi.

1899   The American advertising agency J. Walter Thompson establishes a London office. Publication of *The Theory of the Leisure Class* by Thorstein Veblen.

1900   Universal Exposition in Paris. Publication of *Guide to Paris, the Exhibition and the Assembly* by P. Geddes and S. Dewey.

1901   First wireless transatlantic telegraph transmission. Publication of *L'opinion et la foule* by Gabriel Tarde. First polemics between Lenin and his opponents on the role of the revolutionary press; birth of Agitprop.

1902   Publication of *Mutual Aid* by Peter Kropotkin (compilation of articles published in *Nineteenth Century* between 1890 and 1896). Publication of *Imperialism* by J. A. Hobson.

1903   Publication of *Shop Management* by F. W. Taylor.

1904   The Russian physiologist I. Pavlov receives the Nobel Prize.

1906   Transmission of the human voice by radio (Fessenden). Creation of the International Radiotelegraph Union.

1907   Transmission of a photograph by belinography (Belin). Agreement on the sharing of the world market in electrical industry (General Electric and AEG). Founding of Pathé.

1909   Creation of the first syndicate for the distribution of comic strips, crossword puzzles, and other features.

1910   Beginning of the star system (Mary Pickford). First applications of Fordism in Ford workshops.

1911   First marketing firms in the United States.

1913   Hollywood separates from Los Angeles.

1914   Inauguration of the Panama Canal. French films dominate the European market. French law on the "repression of press indiscretions in time of war." Strike against Taylorist methods at the Renault auto works. Publication of *Behavior: An Introduction to Comprehensive Psychology* by J. B. Watson.

1915   Creation by the French government of the *section d'information*. Founding of the King Feature Syndicate and beginning of the internationalization of comics ("Bringing up Father").

1916   Collapse of French film production. Germany protects its film market by regulating imports. Publication of *Imperialism, the Highest Stage of Capitalism* by V. I. Lenin.

1917   Founding of UFA (Universum-Film-Aktiengesellschaft), basis of postwar German film leadership. Creation in the United States of the Committee on Public Information (Creel Committee) and passage of the Espionage Act, followed in 1918 by the Sedition Act. Founding of the Four A's (American Association of Advertising Agencies). Publication of *The Coming Polity* by P. Geddes and V. Banford.

1918   Establishment in London of Crewe House, entrusted to Lord Northcliffe. Creation in France of a special committee for "aesthetic propaganda abroad." Publication of *The Polish Peasant in Europe and America* by W. Thomas and F. Znoniecki, members of the "Chicago School."

1919   World hegemony of U.S. cinema. Acquisition of the British subsidiary of Marconi by General Electric at the instigation of the U.S. Navy. Founding in Moscow of the Third International.

1920   Publication in Weimar Germany of the first illustrated news magazines. Creation of the Radio Corporation of America (RCA). Introduction of the management system at General Motors.

1921   Third congress of the Communist International: the *Komintern* becomes an instrument of international communication.

1922   First regular radio broadcasts. The Soviets restore *Glavlit*, censorship commission. Behaviorist psychology (J. B. Watson) comes into prominence in advertising industry. Rise of Fordism.

1924   Hollerith Tabulating Machines becomes International Business Machines (IBM). Catholic missionary press totals 411 magazines in different languages.

1925   AT&T/Western Electric cedes its international network of factories in telephone manufacture to International Telephone and Telegraph, (ITT), founded in 1920. Rise of marketing studies in the United States. Foundation of the International Union of Radiodiffusion (IUR).

1926   Beginning of sound cinema. RCA sets up the United States' first national radio network, NBC. In England, radio becomes a public service (BBC). Creation of the Empire Marketing Board and the British Council. Publication in Madrid of the articles of Spanish philosopher Ortega y Gasset on the rise of technical culture under American domination.

1927   Beginning of U.S. empiricist sociology of mass communications (*Propaganda Technique in the World War* by Harold Lasswell). The Sacco-Vanzetti Affair. Establishment of the first two international advertising networks (J. Walter Thompson and McCann-Erickson).

1928   First steps in the theory of public relations (E. Bernays and Ivy Lee).

1929   First regular Soviet radio broadcasts destined for abroad. First experimental television broadcast in Great Britain (John Logie Baird). Procter & Gamble sets up the first internal market research department.

1930   First experiments with television in the United States (V. Zworykin). Paris Accord on the sharing of the world market in sound cinema apparatus. Bertolt Brecht publishes his "Theory of Radio." Concentration of film production in the hands of the five majors.

1932   Repression and tightening of censorship in U.S.S.R. (A. Zdanov). Creation by the BBC of the Empire Service. Merger of the International Telegraph Union and the International Radiotelegraph Union to form the International Union of Telecommunications (IUT).

1933   Hitler comes to power; creation of the Ministry of Propaganda and Enlightenment of the People (Goebbels). Election of Franklin Roosevelt and the beginning of the New Deal. Publication of *The Human Problems of an Industrial Civilization* by Elton Mayo. Publication of *Massenpsychologie des Faschismus* by Wilhelm Reich.

1934   Creation of the Federal Communications Commission (FCC). Publication of *Technics and Civilization* by Lewis Mumford. Publication of "Americanism and Fordism" by Antonio Gramsci.

1936   First issue of *Life*. First Gallup polls in a political campaign. Inauguration of BBC Television Studios.

1937   Publication in the United States of *The Spirit and Structure of German Fascism* by Robert A. Brady and of *The Structure of Social Action* by Talcott Parsons, as well as the first issue of *The Public Opinion Quarterly*. Formulation of the "International Code of Advertising Practice" by the International Chamber of Commerce.

1938   Founding of the French Institute of Public Opinion (IFOP). CBS radio program "The War of the Worlds" by Orson Welles. American riposte to the Nazi offensive in Latin America. Founding of the International Advertising Association in New York.

1939   Publication of the French edition of *The Rape of the Masses* (S. Chakotin). Invention of the Audimeter. Rise of "psychological warfare" concept. First television broadcasts in the United States.

1942   Creation of Voice of America (VOA), the Office of War Information (OWI), and OSS (Office of Strategic Services).

1944   Publication of *The People's Choice* by P. Lazarsfeld, B. Berelson, and H. Gaudet.

1945   Completion of the last great calculator, ENIAC, conceived in military secrecy.

1947   Promulgation of the National Security Act and creation of the CIA. UIT and Universal Postal Union are integrated into the United Nations system. Conference in Atlantic City on the use of the radio spectrum. Creation of GATT (General Agreement on Tariffs and Trade), also becoming part of the United Nations system.

1948   Publication of *Cybernetics: Control and Communication in the Animal and Machine* by Norbert Wiener. Appearance of the concept of "culture industry" (Frankfurt School). Discovery of the transistor (Bardeen, Brattain, and Stockley). The Information Act institutes an Office of International Information (OII) under the State Department.

1949   Formulation of the mathematical theory of information (C. Shannon). Appearance of the notion of "development/underdevelopment" (Point Four of President Truman's State of the Union speech).

1950   First broadcast by the clandestine radio station Radio Free Europe. Decision to build the SAGE system of aerial defense. First cybernetic formulation of the corporation as a system in the United States.

1951   Publication of *The Reds Take a City: The Communist Occupation of Seoul* by Wilbur Schramm and John Riley. Publication of *The Bias of Communication* by the Canadian Harold Innis.

1952   Project for a new discipline of "international communications" as a branch of empiricist sociology. Appearance of the notion of "Third World" (Alfred Sauvy and Georges Balandier).

1953   The U.S. Information Agency (USIA) replaces the OII. First broadcast by the clandestine radio station Radio Liberty.

1954   First transistor radio produced in the United States.

1955   Publication of *Personal Influence* by Paul Lazarsfeld and Elihu Katz. Formulation of theory of the two-step flow.

1956   First radio programs from "Voice of Free Algeria." Publication in France of Roland Barthes's *Mythologies*. First transatlantic underwater telephone cable.

1957   The Soviet Union launches the first artificial satellite, Sputnik. Founding of NASA (National Aeronautics and Space Administration). Publication of *The Political Economy of Growth* by Paul Baran (first formulation of the theory of dependency).

1958   First experiments with the Arpanet computer network. Publication of *The Passing of Traditional Society* by Daniel Lerner.

1959   Publication of studies on intercultural relations by Edward T. Hall (Palo Alto School). IBM perfects the first transistor computer. Geneva conference on the use of the radio spectrum.

1960   Publication of *The Stages of Economic Growth* by W. W. Rostow.

1961   Publication of *Modern Warfare* by the French veteran of Algeria, Colonel Roger Trinquier. Creation of the Alliance for Progress. Publication of *Les damnés de la terre* (The wretched of the earth) by Frantz Fanon with a foreword by Jean-Paul Sartre.

1962   Publication of the report *Social Science Research and National Security* (Lucian Pye and Ithiel de Sola Pool), of *The Diffusion of Innovations* by Everett Rogers, and of *The End of Ideology* by Daniel Bell. Launching of the first telecommunications satellite (Telstar) linking the United States with Europe. Publication of *The Gutenberg Galaxy* by Marshall McLuhan, and of *The Production and Distribution of Knowledge in the United States* by Fritz Machlup.

1963   Venezuela emerges as a new center of critical media theory.

1964   Creation of Intelsat (International Communications Satellite). The U.S.S.R. launches its first communications satellite (Molnya).

1965   Launching of the first geostationary satellite, Early Bird, of the Intelsat system.

1966   French structuralism gains attention as an intellectual current.

1967   Appearance of the notion of *self-reliance*. Publication of *Desarrollo y dependencia en America Latina* (Development and dependency in Latin America) by F. H. Cardoso and Enzo Faletto. First broadcast by Mondovision.

1968   Appearance of the concept of the "postindustrial society" (Daniel Bell). The Cultural Congress held in Havana legitimates the notion of cultural imperialism.

1969   New centers of critical media theory (Chile, Argentina, Great Britain, Scandinavia, United States). Publication of *War and Peace in the Global Village* by Marshall McLuhan and Quentin

Fiore, and of *Between Two Ages* by Zbigniew Brzezinski. Publication of the first works of Paulo Freire.

1970   Publication of *The War of Ideas in Contemporary International Relations* by Soviet author Georgi Arbatov. First disinformation campaign against the constitutional government in Chile. First debates on communication policy in UNESCO. First Japanese report on the information society.

1971   Creation of Intersputnik.

1972   Hearings in the U.S. Senate by the Fulbright Commission on the USIA's government propaganda. Debate in UNESCO and the United Nations General Assembly on an agreement to regulate direct-broadcast satellites.

1973   Beginnings of debate on rights to the sea which will continue throughout the decade. Proposal by nonaligned countries for a "new world economic order" (Algiers). First steps in the move for a "new world information and communication order" (NWICO). First measures by the Japanese government to encourage the international expansion of electronics corporations.

1974   Launching of the SACI tele-education project in Brazil. Debate on remote-sensing satellites in the U.N. Assembly.

1975   Creation of a news agency pool by nonaligned countries. Report of the Trilateral Commission on the "ungovernability" of Western democracies. Beginning of research in France on the "cultural industries."

1976   Hearings in the U.S. Senate by the Church Commission on Foreign and Military Intelligence. SITE project for use of satellites in India for educational purposes. Intergovernmental Conference on communication policy under the auspices of UNESCO at San José, Costa Rica.

1977   Creation of the International Commission for the study of communications problems (UNESCO) under the presidency of Sean McBride. Porat Report on the information society published by the U.S. government.

1978   French report on the computerization of society by Simon Nora and Alain Minc. Adoption of the notion of "cultural industries" by the Council of Europe. Intergovernmental Conference on communication policies in Kuala Lumpur.

1979   World Administrative Radio Conference (WARC) in Geneva (IUT). Publication in France of the Rigaud report on foreign cultural relations and of *La condition postmoderne* (The postmodern condition) by Jean-François Lyotard. Publication of the Clyne Report in Canada.

1980   Publication of the McBride Commission report. Publication of *Arts de faire* (The uses of everyday life) by Michel de Certeau. Entry of the Japanese management model into France. Intergovernmental Conference on communication policies in Yaoundé (Cameroon).

1981   IBM brings out its first personal computer.

1982   The Falklands/Malvinas War: first appearance of the practice of organizing pools of journalists. French strategy of the development of the electronics sectors. Report on technology, employment, and growth presented at the Versailles summit by French President François Mitterrand.

1983   U.S. Senate report on telecommunications. Publication of Theodor Levitt's article on "globalization." Embargo on information by the Pentagon during the intervention on the island of Grenada.

1984   Effective dismantling of AT&T as a result of a 1982 legal decision and deregulation of telecommunications in the United States. Green Paper by the European Community on "Television Without Frontiers." Beginning of EC debate on a directive related to this program and the Council of Europe's convention on transborder television.

1985   Escalation of mergers and takeover bids in the realm of communications, from advertising and publishing to radio and television. Intelsat goes into competition with a private system. The United States withdraws from UNESCO.

1986   Start of negotiations on free trade in services (Uruguay Round of GATT). Britain withdraws from UNESCO.

1987   European Community Green Paper on telecommunications.

1989   Merger of Time and Warner into a multimedia conglomerate. Sony buys Columbia Pictures. The EC Directive and Council of Europe Convention on "Television Without Frontiers" both adopted. Fall of the Berlin Wall and the end of the informational closure of Eastern Europe. With U.S. intervention in Panama, the international role of CNN (Cable News Network) first becomes apparent. Manipulation of information during the fall of the Ceauşescus in Romania.

1990   Gulf Crisis: CNN propelled to the rank of global television network. Matsushita acquires MCA-Universal Studios. The competitive stakes in high-definition television (HDTV) become clear.

1991   The Gulf War: the U.S. Armed Forces institute the pool system for journalists. Negotiations for the limitation of imports of arms and "destabilizing" technologies. Opening of negotiations for NAFTA (North American Free Trade Agreement) for the creation of a free trade zone between Canada, the United States, and Mexico. Failed coup d'état in Moscow and self-determination for the republics of the former Soviet Union.

1992   Process of ratification of the Maastricht Treaty by members of the EEC. Expo '92 Seville. The so-called "double leapfrog" alliance of IBM-Toshiba-Siemens. Fall in profits of Japanese electronic and computer groups. "Global Forum" of nongovernmental organizations (NGO's) in Rio de Janeiro. Operation Restore Hope in Somalia backed by media logistics.

Notes

1. The Emergence of Technical Networks

1. Y. Stourdzé, "Généalogie des télécommunications françaises," in *Les réseaux pensants*, coordinated by Giraud et al., Paris, Masson, 1977.

2. To situate the historical landmarks of the evolution of the telegraph and the telephone, we used two classics: A. Belloc, *La télégraphie historique depuis les temps les plus reculés jusqu'à nos jours,* Paris, Firmin-Didot, 1888; C. Bertho, *Télégraphes et téléphones. De Valmy au microprocesseur,* Paris, Le livre de poche, 1981.

3. See E. Vaillé, *Histoire générale des postes françaises,* Paris, PUF, 1947–55, 6 volumes.

4. B. Delepinne, *Histoire de la poste internationale en Belgique,* Brussels, Presses H. Wellens & W. Godenne, 1952, p. 80.

5. F. Staff, *The Penny Post,* London, Lutterworth Press, 1964, p. 22.

6. S. Maïski quoted in J. Medvedev, *Le secret de la correspondance est garanti par la loi,* Paris, Julliard, 1972, p. 121.

7. G. Freund, *Photographie et société,* Paris, Seuil, 1974.

8. On the horrors of the Crimean War, one of the most eloquent texts is Leo Tolstoy's *Sebastopol Sketches,* trans. D. McDuff, London, Penguin, 1986, written in 1855–56.

9. P. Leroy-Beaulieu in *La grande encyclopédie,* Paris, 1902.

10. M. Barbance, *Histoire de la Compagnie Générale Transatlantique,* Paris, Arts et métiers graphiques, 1955.

11. "The Sugar Cane," quoted in M. Moreno Fraginals, *El ingenio, complejo económico social cubano del azúcar,* Havana, Editorial de Ciencias Sociales, 1978.

12. E. Hobsbawm, *The Age of Empire (1875–1914),* London, Weidenfeld & Nicolson, 1987.

13. N. R. Danielian, *A.T.& T.: The Story of Industrial Conquest,* New York, Vanguard Press, 1939.

14. See the dossier "Entre guerre et paix" ("Between War and Peace") coordinated by A. Lefébure, *Interférences,* no. 1, Paris, 1981.

15. Information published by the Bureau international des administrations télégraphiques, report, Bern, 1890, 3d edition.

16. Bertho, *Télégraphes et téléphones.*

17. P. Bata, "Les câbles sous-marins des origines à nos jours," *Télécommunications,* no. 45, Paris, October 1982.

18. M. Palmer, *Des petits journaux aux grandes agences. Naissance du journalisme moderne,* Paris, Aubier, 1983, p. 194.

19. See J. O. Boyd and M. Palmer, *Le trafic des nouvelles. Les agences mondiales d'information,* Paris, Alain Moreau, 1981.

20. Palmer, *Des petits journaux.*

21. R. Gubern, *El lenguage de los comics,* Barcelona, Ediciones Peninsula, 1974, p. 70.

22. P. Paranagua, *Cinema na America latina,* Porto Alegre, L & PM, 1984.

23. Quoted in J. M. Frodon, "L'Amérique et ses démons," *Le Monde,* May 7, 1992, p. 30.

24. P. Bachlin, *Histoire économique du cinéma,* Paris, La Nouvelle Edition, 1947.

25. G. Kennan, *American Diplomacy 1900–1950,* London, Secker & Warburg, 1952.

26. F. Williams, *The Right to Know,* London, Longman, 1969.

27. On the history of United Fruit Company, by one of its former employees, see T. Mc-Cann, *An American Company: The Tragedy of United Fruit,* New York, Crown, 1976. For a critical history containing insightful analyses of the company's communications system, see C. Fontanellas, ed., *United Fruit Co.,* Havana, Editorial de Ciencias Sociales, 1976.

28. H. Peyret, *Histoire des chemins de fer en France et dans le monde,* Paris, Société d'éditions françaises et internationales, 1949.

29. *Les archives diplomatiques,* Paris, 1888.

30. P. Virilio, "L'empire de l'emprise," *Traverses,* no. 13, December 1978. See also, by the same author, *Speed and Politics,* New York, Semiotext(e), 1986; and in collaboration with S. Lotringer, *Pure War,* New York, Semiotext(e), 1983.

31. M. Ferro, "Images de l'histoire," *Traverses*, no. 13, p. 52.

32. A. Mattelart and H. Schmucler, *Communication and Information Technologies: Freedom of Choice for Latin America?,* trans. D. Buxton, Norwood, N.J., Ablex, 1985.

33. P. Virilio, "L'empire."

34. Quoted in ibid. These remarks by the French engineer could be generalized. See, for example, the analyses of historian A. D. Chandler on the relation between the formation of managerial philosophy and the implantation of the railroad system in the United States: A. D. Chandler, *The Visible Hand: The Managerial Revolution in American Business,* Cambridge, Mass., Harvard University Press, 1977.

35. P. Lafargue, *Le droit à la paresse,* Paris, Maspero, 1976.

36. F. W. Taylor, *Principles and Methods of Scientific Management,* New York, Harper, 1911. This work had been preceded by several articles, including "Shop Management" (1903), published in *Transactions,* the organ of the American Society of Mechanical Engineers.

37. E. Pouget, *L'organisation du surmenage: Le système Taylor,* Paris, Librairie des sciences politiques et sociales, M. Rivière, 1914.

38. Ibid.

39. 1909 JWT Blue Book, 1909–10, original edition reproduced in facsimile for the centennial of the agency. *Advertising Age,* December 7, 1964.

40. T. Schieder, "Political and Social Developments in Europe," *The New Cambridge Modern History,* vol. 11, pp. 253–54. See also the pioneering work of W. Sombart, *Der moderne Kapitalismus,* Munich and Leipzig, 1902–1928, 3 vols.

41. G. Gérault, *Les expositions universelles envisagées au point de vue de leurs résultats économiques,* Paris, Librairie société du recueil général des lois et des arrêts, 1902, p. 23.

42. Ibid.

43. F. Maquaire et al., *Rapport des délégués mécaniciens en précision à l'Exposition universelle de Philadelphie* (1876), Paris, Imprimerie nationale, 1879, p. 21.

44. E. Zola, *L'argent*, Paris, 1891.

45. M. R. Trouillot, "Good Day Columbus: Silences, Power and Public History," *Public Culture* 3, no. 1, Fall 1990, p. 15.

46. Exposition Universelle of 1867 in Paris. The reports of the international jury are published by M. Chevalier, ed., Paris, Imprimerie administrative de Paul Dupont, 1868, p. cdxc.

47. Ibid., p. vi.

48. F. Nietzsche, *The Complete Works of Friedrich Nietzsche,* vol. 2, ed. Oscar Levy, trans. A. Collins, London, Russell & Russell, 1909, p. 39.

49. J. A. Hobson, *Imperialism: A Study,* Ann Arbor, University of Michigan, 1988, p. 215 (original edition 1902). On "the national profile" in the expositions, see P. Greenhalgh, *Ephemeral Vistas: The Expositions Universelles, Great Exhibitions and World's Fairs, 1851–1939,* Manchester, Manchester University Press, 1988.

50. D. McKie, "The World Economy: Interdependence and Planning," *The New Cambridge Modern History,* vol. 12, chap. 3, p. 43.

2. The Age of Multitudes

1. C. H. Cooley, *Social Organization*, New York, Charles Scribner's Sons, 1901, p. 65.

2. Ibid., p. 61. See also, by the same author, *Sociological Theory and Social Research,* New York, Henry Holt, 1930 (with an introduction by R. Cooley Angell).

3. On the work of C. H. Cooley, see L. A. Coser, *Masters of Sociological Thought,* New York, Harcourt Brace Jovanovich, 1971; L. Belman, "The Idea of Communication in the Work of Charles Horton Cooley," *Journal of Communication Inquiry* 1, no. 2, Spring 1975.

4. L. Belman, ibid.

5. J. Carey and J. Quirk, "The History of the Future," in *Communications Technology and Social Policy,* ed. G. Gerbner et al., New York, John Wiley, 1973.

6. M. Chevalier, *Le système de la Méditerranée*, Paris, Le Globe, 1832, p. 47. On the notion of network (*réseau*) see P. Musso, "Aux origines du concept moderne: Corps et réseau dans la philosophie de Saint-Simon," *Quaderni*, no. 3, Paris, 1988.

7. A. Comte, *Cours de philosophie positive*, Paris, 1830–42.

8. H. Spencer, *The Principles of Sociology*, New York, D. Appleton, 1898. The author produced the four volumes of this work between 1876 and 1896.

9. See the analyses of M. L. De Fleur, *Theories of Mass Communications*, New York, D. McKay, 1966.

10. T. Veblen, *The Theory of the Leisure Class,* New York, Modern Library, 1943 (original edition 1899).

11. R. K. Merton, "Manifest and Latent Functions: Toward a Codification of Functional Analysis in Sociology," in *Social Theory and Social Structure*, Glencoe, Ill., Free Press, 1951, 2d edition.

12. T. Adorno, "Veblen's Attack on Culture" (1941), in *Prisms,* Cambridge, Mass., MIT Press, 1981.

13. F. Ratzel, *Politische Geographie*, Munich, Oldenburg, 1897. For exegeses of this work, see A. L. Sanguin, "En relisant Ratzel," *Annales de Géographie,* September–October 1990; and G. Mercier, "Le concept de propriété dans la géographie politique de Ratzel," ibid.

14. R. Brunet, "Usage de l'espace," *Non!*, Paris, July–August 1981.

15. A. T. Mahan, quoted in *American Imperialism and the Philippine Insurrection*. Selections from *Congressional Hearings* (1902), ed. H. F. Graff, Boston, Little, Brown, 1969, p. ix.

16. Ibid., p. viii.

17. E. Reclus, *Nouvelle géographie universelle*, Paris, Librairie Hachette, 1894, vol. 19, pp. 794–95. See also M. Fleming, *The Geography of Freedom: The Odyssey of Elisée Reclus*, introduction by G. Woodcock, Montreal-New York, Black Rose Books, 1985. By P. Kropotkin, see in particular *Mutual Aid*, introduction by G. Woodcock, Montreal-New York, Black Rose Books, 1984.

18. F. Tönnies, *Community and Society*, trans. C. P. Loomis, East Lansing, Michigan State University Press, 1957. In the original German: *Gemeinschaft und Gesellschaft*, Stuttgart, 1887.

19. G. Le Bon, *Psychologie des foules*, Paris, Alcan, 1895. (In American translation: *The Crowd: A Study of the Popular Mind*, New York, Viking/Compass, 1966.) See also S. Barrows, *Distorting Mirrors: Visions of the Crowd in the Late Nineteenth-Century France*, New Haven, Yale University Press, 1981; S. Moscovici, *The Age of the Crowd: A Historical Treatise on Mass Psychology*, Cambridge, Cambridge University Press, 1985, trans. from French.

20. See the dossier devoted to the work of G. Simmel published in *Les Cahiers du Grif*, Paris, Editions Tierce, no. 40, Spring 1989.

21. T. R. Malthus, *An Essay on the Principle of Population* (1798), London, Dent/Everyman's Library, 1973.

22. A. Mattelart, "Une lecture idéologique de l'Essai sur le principe de population," *L'Homme et la Société*, no. 15, January–March 1970.

23. T. Parsons, *The Structure of Social Action*, New York, McGraw-Hill, 1937.

24. G. Tarde, *The Laws of Imitation*, trans. E. C. Parsons, New York, Henry Holt, 1903. First French edition, 1890.

25. K. Lang and G. E. Lang, "The 'New' Rhetoric of Mass Communication Research: A Longer View," *Journal of Communication* 33, no. 3, Summer 1983.

26. R. A. Nye, *The Origins of Crowd Psychology: Gustave Le Bon and the Crisis of Mass Democracy in the Third Republic*, London, Sage, 1975. See also the thesis of J. Van Ginneken, *Crowds: Psychology and Politics 1871–1899*, Amsterdam, University of Amsterdam, 1989.

27. Ministère du commerce, de l'industrie, des postes et des télégraphes, Exposition internationale de Chicago en 1893. Reports published by M. C. Krantz (ed.), *Congrès tenu à Chicago en 1893*, Paris, Imprimerie nationale, 1894.

28. Exposition universelle de 1900, international feminist congress held at the Palais des Congrès. Report presented by Madame Vincent: "Le travail des bonnes," Paris, 1900.

29. E. Monod, *L'exposition universelle de 1889*, Paris, E. Dentu, 1890, vol. 2, p. 283.

30. K. K. Mumtaz, *Art and Imperialism: The Impact of British Rule on the Arts and Crafts in India*, Pakistan Study Group, Monograph Series no. 1, 1980. See also W. G. Archer, *India and Modern Art*, London, 1959.

31. G. Gérault, *Les expositions universelles envisagées au point de vue de leurs résultats économiques*, Paris, Librarie société du recueil général des lois et des arrêts, 1902, p. 204.

32. W. Benjamin, *Paris, capitale du XIXème siècle. Le livre des passages*, Paris, Le Cerf, 1989.

33. Rastignac, "Courrier de Paris," *L'Illustration*, no. 2411, May 11, 1889, p. 394.

34. For an analysis of the theme of communication in the works of Karl Marx, see Y. de La Haye, ed., *Marx & Engels on the Means of Communication: A Selection of Texts*, New York, International General, 1980.

35. G. Balandier, *Le désordre*, Paris, Fayard, 1988, p. 228.

36. Voir C. Weill, *L'Internationale et l'Autre*, Paris, Arcantère, 1987.

37. C. De Paepe, *Les services publics*, Brussels, Bibliothèque populaire, J. Milot Editeur, 1895, p. 22.

38. Ibid., p. 11.

39. Ibid.

40. Ibid., p. 27.

41. Ibid., p. 146.

42. A. M. Thiesse, *Le roman du quotidien. Lecteurs et lectures populaires à la Belle Epoque*, Paris, Le Chemin Vert, 1984, p. 117.

43. Ibid., pp. 118–19.

44. V. I. Lenin, "What Is to Be Done? (1901–1902)," *Lenin about the Press*, Prague, International Organization of Journalists, 1972, pp. 95–96.

3. The Invisible Management of the Great Society

1. W. Benjamin, "Théories du fascisme allemand. A propos du recueil de textes 'La Guerre et les Guerriers' de Ernst Jünger," trans. and published by *Interférences*, no. 1, 1981, p. 28.

2. G. Sylvester Viereck, *Spreading Germs of Hate*, New York, Horace Liveright, 1930, pp. 153–54.

3. G. G. Bruntz, "Allied Propaganda and the Collapse of German Morale in 1918," *Public Opinion Quarterly* 2, 1938, reprinted in *A Psychological Warfare Casebook*, ed. W. E. Daugherty and M. Janowitz, published for Operations Research Office, Baltimore, Johns Hopkins University Press, 1958, p. 101.

4. P. Von Hindenburg, *Out of My Life*, New York, Harper, 1921. Trans. from German.

5. Bruntz, "Allied Propaganda and the Collapse of German Morale in 1918."

6. Quoted by P. Bachlin, *Histoire économique du cinéma*, Paris, La Nouvelle Édition, 1947, pp. 32–33.

7. J. R. Mock and C. Larsen, *Words That Won the War: The Story of the Committee on Public Information 1917–19*, Princeton, Princeton University Press, 1939.

8. E. E. Dennis et al., *The Media at War: The Press and the Persian Gulf Conflict. A Report of the Gannett Foundation*, New York, Columbia University, 1991.

9. M. Balfour, *Propaganda in War 1939–1945. Organisation, Policies and Publics in Britain and Germany*, London, Routledge & Kegan Paul, 1979.

10. Quoted by A. J. P. Taylor, *English History 1914–45*, Oxford, Oxford University Press, 1965, p. 26.

11. The information about censorship and how it was organized during World War I is taken from press historian P. Albert's contribution to the *Histoire générale de la presse française de 1871 à 1940*, Paris, Bellanger, PUF, 1972, vol. 3.

12. G. Le Bon, *Enseignements psychologiques de la guerre européenne*, Paris, Ernest Flammarion, 1916, p. 320.

13. H. Blankenhorn, *Adventures in Propaganda: Letters from an Intelligence Officer in France*, Boston, 1919.

14. F. Baldensperger, *Note sur les moyens d'action intellectuelle de la France à l'étranger*, Paris, Imprimerie L. De Matteis, 1917.

15. Ibid., p. 61.

16. Several texts illustrating this period have been included in *Communication and*

Class Struggle, an anthology in two volumes, ed. A. Mattelart and S. Siegelaub, New York, International General, 1979 and 1983.

17. B. Brecht, "Radio as a Means of Communication," *Screen* 20, no. 3/4, Winter 1979–80. On Brecht's contribution to the theory of the media, see H. M. Enzensberger, "Constituents of a Theory of the Media," in *The Consciousness Industry: On Literature, Politics and the Media,* New York, Seabury Press, 1974.

18. G. Freund, *Photographie et société,* Paris, Seuil, 1974.

19. M. Balfour, *Propaganda in War.*

20. N. J. Spykman, *America's Strategy in World Politics: The United States and the Balance of Power,* New York, Harcourt, Brace & World, 1942, p. 233.

21. J. Aronson, *The Press and the Cold War,* New York, Bobbs-Merrill, 1970, p. 25.

22. *History of Communications: Electronics in the United States Navy,* Washington, D.C., Government Printing Office, 1963.

23. For a chronology, see *Electrical Communication* 46, no. 4, London, 1971.

24. For Great Britain, see L. Denny, *America Conquers Britain: A Record of Economic War,* New York, 1930. For France, see C. Bertho, *Télégraphes et téléphones. De Valmy au microprocesseur,* Paris, Le livre de poche, 1981.

25. A. Sampson, *The Sovereign State of ITT,* New York, Stein & Day, 1973.

26. Bachlin, *Histoire économique du cinéma.*

27. Balfour, *Propaganda on War,* chap. 1 ("The Demythologising of Crewe House").

28. H. D. Lasswell, *Propaganda Techniques in the World War,* New York, Alfred Knopf, 1927, p. 14.

29. Ibid., pp. 220–21.

30. R. E. Park, "News and Opinion," reprinted in *The Collected Papers of R. E. Park,* Glencoe, Ill., Free Press, 1955. See also the author's doctoral thesis, presented in Germany in 1904, translated and published only in 1972, under the title *The Crowd and the Public,* Chicago, University of Chicago Press, 1972.

31. W. McDougall, *An Introduction to Social Psychology,* Boston, Luce, 1908; *The Group Mind,* London, Cambridge University Press, 1920. See also M. Conway, *The Crowd in Peace and War,* London, Longman, Green, 1916. On the relationship between American authors of the time and European social psychology, see J. Van Ginneken, *Crowds: Psychology and Politics 1871–1899,* Amsterdam, University of Amsterdam, 1989.

32. J. Rassak, *Psychologie de l'opinion et de la propagande politique,* Paris, Librairie des sciences politiques et sociales, Marcel Rivière Editeur, 1927.

33. L. Richard, *Le nazisme et la culture,* Paris, François Maspero, 1978. On the organization of the film industry, see D. Stewart Hull, *Film in the Third Reich: Art and Propaganda in Nazi Germany,* Berkeley, University of California Press, 1969.

34. S. Tchakhotine, *Le viol des foules par la propagande politique,* Paris, Gallimard, 1952 (original edition 1939). English translation of the original edition: S. Chakotin, *The Rape of the Masses: The Psychology of Totalitarian Political Propaganda,* New York, Alliance / London, Routledge, 1940.

35. R. A. Brady, *The Spirit and Structure of German Fascism,* London, Gollancz, 1937. Neither S. Chakotin nor R. A. Brady refers in their work to Wilhelm Reich (see *The Mass Psychology of Fascism,* New York, Farrar, Straus & Giroux, 1971).

36. P. Reiwald, *De l'esprit des masses,* Geneva, Delachaux et Niestlé, 1946.

37. Tchakhotine, *Le viol,* p. 337.

38. F. Champarnaud, *Révolution et contre-révolution culturelle en URSS. De Lénine à Jdanov,* Paris, Anthropos, 1975.

39. E. Bernays, *Propaganda,* New York, 1928. See also *Crystallizing Public Opinion,*

New York, 1923; *The Engineering of Consent,* ed. E. Bernays, Norman, University of Oklahoma Press, 1955.

40. See S. Ewen, *Captains of Consciousness: Advertising and the Social Roots of the Consumer Culture,* New York, McGraw-Hill, 1976.

41. D. Bell, "Modernity and Mass Society: Diversity of Cultural Experience," published in French under the title of "Les formes de l'expérience culturelle," *Communications,* no. 2, 1963.

42. W. I. Thomas and F. Znaniecki, *The Polish Peasant in Europe and America,* Chicago, University of Chicago Press, 1918–21, 5 volumes.

43. G. W. Allport, "Attitudes," in *Handbook of Psychology,* ed. C. Murchison, Worcester, Mass., Clark University Press, 1965. By the same author, in collaboration with L. Postman, see *The Psychology of Rumor,* New York, H. Holt, 1947.

44. E. Mayo, *The Human Problems of an Industrial Civilization,* New York, Macmillan, 1933.

45. T. Parsons, *The Structure of Social Action,* New York, McGraw-Hill, 1937.

46. C. Wright, *Mass Communication: A Sociological Perspective,* New York, Random House, 1959; R. K. Merton, *Social Theory and Social Structure,* Glencoe, Ill., Free Press, 1951, 2d edition.

47. T. Parsons, *Sociological Theory and Modern Society,* Glencoe, Ill., Free Press, 1962.

48. J.-M. Vincent, "La sociologie en contrepoint," *L'Homme et la Société,* no. 3, 1990, p. 47.

49. See J. Schultze, "Professionalism in Advertising: The Origin of Ethical Code," *Journal of Communication* 31, no. 2, Spring 1981; E. Clarke, *The Want-Makers,* London, Hodder & Stoughton, 1988.

50. F. Morris et al., *A History of the People America,* New York, Rand McNally, 1971.

51. H. Cantril et al., *The Invasion from Mars,* Princeton, Princeton University Press, 1940.

52. A. Gramsci, *Letteratura e vita nazionale,* Rome, Riuniti, 1977, Part III.

53. A. Gramsci, "National-Popular Literature; The Popular Novel, and Observations on Folklore" in *Communication and Class Struggle,* vol. 2, ed. A. Mattelart and S. Siegelaub, New York, International General, 1983, p. 73.

54. A. Gramsci, *Quaderni del carcere,* Turin, Einaudi, 1975, quaderno n. 13, §18.

55. A. Gramsci, "Americanismo e fordismo," *Note sul Machiavelli sulla politica e sullo stato moderno,* Rome, Riuniti, 1977.

56. L. Pirandello interviewed by C. Alvaro, *L'Italia Letteraria,* April 14, 1929. Quoted by A. Gramsci in ibid., Part III.

57. J. C. Mariátegui, *Siete ensayos de interpretación de la realidad peruana,* Lima, Imprenta Amauta, 1928.

58. J. Ortega y Gasset, *La Rebelion de las masas,* Madrid, *Revista de Occidente* / Alianza Editorial, 1983, pp. 192–93. (English translation: *The Revolt of the Masses,* New York, Norton, 1932). See also R. Aron and A. Dandieu, *Le cancer américain,* Paris, Editions Rieder, 1931; A. Siegfried, *La crise de l'Europe,* Paris, Calmann-Levy, 1935.

4. The Shock of Ideologies

1. N. J. Spykman, *America's Strategy in World Politics,* New York, Harcourt, Brace & World, 1942.

2. J. Hale, *Radio Power Propaganda and International Broadcasting,* London, Elek Books, 1975.

3. A. J. Tudesq, *La radio en Afrique noire,* Paris, Pédone, 1983.

4. P. Miquel, *Histoire de la radio et de la télévision*, Paris, Richelieu, 1973.

5. A. L. Woll, "Hollywood's Good Neighbor Policy: The Latin Image in American Film, 1939–1946," *Journal of Popular Film* 3, no. 4, 1974.

6. R. Schickel, *The Disney Version: The Life, Times, Art and Commerce of Walt Disney*, New York, Discus Books/Avon, 1969.

7. P. Mélandri, *Histoire des Etats-Unis depuis 1865*, Paris, Nathan, 1984.

8. C. Layton, *L'Europe et les investissements américains*, Paris, Gallimard, 1968.

9. Spykman, *America's Strategy*, p. 33.

10. H. D. Lasswell, "Political and Psychological Warfare," in *Propaganda in War and Crisis*, ed. D. Lerner, New York, George W. Stewart , 1950. By the same author: *World Revolutionary Propaganda*, New York, Alfred Knopf, 1939.

11. M. Balfour, *Propaganda in War 1939–1945*, London, Routledge & Kegan Paul, 1979.

12. L. Farago, *War of Wits: The Anatomy of Espionage and Intelligence*, Funk & Wagnalls, 1954, p. 323. See also ed. Farago, *German Psychological Warfare: A Critical, Annotated and Comprehensive Survey and Bibliography*, New York, G. P. Putnam's, 1941. Another enduring classic is P. M. A. Linebarger, *Psychological Warfare*, Washington, D.C., Infantry Journal Press, 1948.

13. C. Kluckhohn, *Mirror for Man*, New York, McGraw-Hill, 1949.

14. E. A. Shils and M. Janowitz, "Cohesion and Disintegration in the Wehrmacht in World War II," *Public Opinion Quarterly* 12, 1948. See also a classic on formal organizations: S. A. Stouffer et al., eds., *The American Soldier: Studies in Social Psychology in World War II*, Princeton, Princeton University Press, 1949.

15. J. W. Riley and L. S. Cottrell, "Research for Psychological Warfare," *Public Opinion Quarterly* 21, 1957.

16. C. I. Hovland, A. A. Lumsdaine, and F. D. Sheffield, *Experiments on Mass Communication*, Princeton, Princeton University Press, 1949.

17. P. F. Lazarsfeld, "The Prognosis for International Communication Research," *Public Opinion Quarterly* 16, 1953, p. 482.

18. W. Dizard, *The Strategy of Truth: The Story of the U.S. Information Service*, Washington, D.C., Public Affairs Press, 1961.

19. C. Moisy, *L'Amérique sous les armes*, Paris, Le Seuil, 1971.

20. L. S. Rodberg and D. Shearer, eds., *The Pentagon Watchers*, New York, Doubleday Anchor, 1970.

21. B. J. Williams, "The Importance of Research and Development to National Security," *Military Review*, February 1950, p. 11.

22. L. S. Cottrell, "Psychological Warfare: A Misnomer," in *A Psychological Warfare Casebook*, ed. W. E. Daugherty and M. Janowitz, Baltimore, Johns Hopkins University Press, 1958, pp. 18–19.

23. R. I. Perusse, "Psychological Warfare Reappraised," in ibid.

24. M. Dyer, *The Weapon on the Wall: Rethinking Psychological Warfare*, Baltimore, Johns Hopkins University Press, 1959, p. 32.

25. P. F. Lazarsfeld, B. Berelson, and H. Gaudet, *The People's Choice*, New York, Duell, Sloan & Pearce, 1944.

26. E. Katz and P. F. Lazarsfeld, *Personal Influence: The Part Played by People in the Flow of Mass Communications*, Glencoe, Ill., Free Press, 1970 (1st edition 1955).

27. E. Katz, "The Two-Step Flow of Communication," *Public Opinion Quarterly* 21, 1957.

28. B. Berelson, "Communications and Public Opinions," in *Mass Communications*, ed. W. Schramm, Champaign-Urbana, University of Illinois Press, 1949, p. 500.

29. A. Inkeles, "The Soviet Characterization of the Voice of America," *Journal of International Affairs*, no. 5, 1951, p. 44. See also his major work *Public Opinion in Soviet Russia*, Cambridge, Mass., Harvard University Press, 1950.

30. J. T. Klapper and L. Lowenthal, "The Contributions of Opinion Research to the Evaluation of Psychological Warfare," *Public Opinion Quarterly* 15, 1951–52, p. 651.

31. W. Schramm et al., *The Nature of Psychological Warfare*, Baltimore, Johns Hopkins University Press, 1958, quoted in Dyer, *Weapon on the Wall*, p. 58.

32. P. F. Lazarsfeld, "The Prognosis for International Communication Research."

33. C. Wright Mills, *The Sociological Imagination*, Harmondsworth, Penguin, 1970, p. 92 (original edition 1959).

34. Quoted in Dyer, *Weapon on the Wall*, p. 25.

35. OECD, *Allocations de ressources dans le domaine de l'informatique et des télécommunications*, Part III, Paris, 1975. On the history of computer science, see N. Metropolis et al., eds., *A History of Computing in the Twentieth Century*, London, Academic Press, 1980.

36. Philippe Breton, *Une histoire de l'informatique*, Paris, La Découverte, 1987, pp. 129–30.

37. J. Fletcher, "Toward Corporate Continuity in Space: The Case for NASA's Future," *Finance*, April 1972.

38. J. E. Webb, *Space: The New Frontier*, Washington, D.C., U.S. Government Printing Office, 1967.

39. A. Mattelart, *Multinational Corporations and the Control of Culture*, Sussex, Harvester Press; Atlantic Highlands, N.J., Humanities Press, 1979, Chapter 3: "The Diffusion of Space Technology."

40. W. Schramm, *Satellites de télécommunications pour l'éducation, la science et la culture*, Paris, UNESCO, 1968 (Collection Etudes et documents d'information, no. 53).

41. Fletcher, "Toward Corporate Continuity in Space."

42. Quoted in J. Woodmansee, *The G.E. Project: The World of a Giant Corporation*, Washington, D.C., North Country Press, 1975.

43. See M. Mattelart, "Education, Television and Mass Culture: Reflections on Research into Innovation," in *Television in Transition*, ed. P. Drummond and R. Paterson, London, BFI, 1985.

44. *Department of State Newsletter*, February 1971.

45. G. Bateson et al., *La nouvelle communication*, presentation by Y. Winkin, Paris, Le Seuil, 1981.

46. E. T. Hall, *The Silent Language*, New York, Doubleday, 1959.

47. E. T. Hall, "The Silent Language in Overseas Business," *Harvard Business Review*, May–June 1960.

48. J. Solomon, "U.S. Managers Remain Focused on Home," *Wall Street Journal* (European edition), July 18, 1989, p. 1.

5. The School of Ruse

1. R. Aron, *Penser la guerre, Clausewitz*, Paris, NRF-Gallimard, 1976, vol. 2, p. 115.

2. G. Deleuze and F. Guattari, *Mille plateaux. Capitalisme et schizophrénie*, Paris, Minuit, 1980 (*A Thousand Plateaus: Capitalism and Schizophrenia*, trans. B. Massumi, Minneapolis, University of Minnesota Press).

3. G. Chaliand, *Stratégies de la guérilla*, Paris, Mazarine, 1979.

4. See Mao Tse-tung, *On Protracted War*, Peking, Foreign Languages Press, 3d edition, 1966. Another classic is the work by Ernesto ("Che") Guevara, founder of Radio Re-

belde, the radio station of the Castroist guerilla movement, which played a key role in the overthrow of the Batista dictatorship in 1959: *Guerrilla Warfare,* New York, Monthly Review Press, 1961. See also Vo Nguyen Giap, *People's War, People's Army,* New York, Praeger, 1962.

5. J. M. Noyer, "De la notion de guérilla à la notion de techno-guérilla. Evolution technologique et transformation des machines de guerre," *Etudes Internationales* 21, no. 2, Université Laval, Québec, June 1990, p. 296.

6. R. Trinquier, *La guerre moderne,* Paris, La Table ronde, 1961, pp. 15–16.

7. F. Fanon, "This Is the Voice of Algeria" (1959), in *Communication and Class Struggle,* vol. 2, ed. A. Mattelart and S. Siegelaub, New York, International General, 1983, p. 211.

8. D. A. Starry, "La guerre révolutionnaire," *Military Review,* February 1967. See also J. Steward Ambler, *The French Army in Politics,* Columbus, Ohio State University Press, 1966.

9. G. Kelly, "Revolutionary War and Psychological Action," *Military Review,* October 1960.

10. General Nemo, "La guerre dans la foule," *Revue de Défense Nationale,* June 1956.

11. R. Trinquier, *Modern Warfare: A French View of Counterinsurgency,* New York, Praeger, 1964.

12. Ibid., p. 26.

13. Ibid., p. 48.

14. F. Kitson, *Low Intensity Operations: Subversion, Insurgency, Peacekeeping,* London, Faber and Faber, 1971.

15. P. Schlesinger, G. Murdock, and P. Elliott, *Televising Terrorism: Political Violence in Popular Culture,* London, Comedia, 1983, p. 166. Also by P. Schlesinger, see "On the Shape and Scope of Counter-Insurgency Thought," in *Power and the State,* ed. G. Littlejohn et al., London, Croom Helm, 1978.

16. General Golbery do Couto e Silva, *Geopolitica do Brasil,* Rio de Janeiro, Livraria José Olympio, 1967 (2d edition). For an analysis of Latin American military elites and their ideology, see the monographic issue devoted to the question by the French journal *Critique* (August–September 1977). Aside from historical articles, there are also analyses of the major geopolitical treatises written by the generals of the South American continent.

17. M. and A. Mattelart, *The Carnival of Images: Brazilian Television Fiction,* New York, Bergin & Garvey (Greenwood Press), 1990, pp. 30–31.

18. Trinquier, *Modern Warfare,* p. 105.

19. Starry, "La guerre révolutionnaire."

20. M. T. Klare, *War without End: American Planning for the Next Vietnams,* New York, Vintage, 1972, p. 90. See also C. Brightman and M. T. Klare, "Social Research and Counterinsurgency," *NACLA Newsletter,* New York-Berkeley, March 1970.

21. M. T. Klare, ibid., p. 91.

22. See also I. L. Horowitz, ed., *The Rise and Fall of Project Camelot,* Cambridge, Mass., MIT Press, 1967.

23. I. de Sola Pool, *The Technologies of Freedom,* Cambridge, Mass., Harvard University Press, 1983.

24. G. Piel quoted in Klare, *War without End,* p. 76. See also P. Dickson, *Think Tanks,* New York, Ballantine, 1971.

25. S. P. Huntington, in *The Crisis of Democracy: Report on the Governability of the Democracies to the Trilateral Commission* (by M. Crozier, S. P. Huntington, and J. Watanuki), New York, New York University Press, 1975, pp. 98–99.

26. J. W. Fulbright, *The Pentagon Propaganda Machine,* New York, Vintage, 1971 (2d edition), p. 157.

27. See J. Aronson, *The Press and the Cold War,* New York, Bobbs-Merrill, 1970.

28. U.S. Senate, *USIA Appropriations Authorization, Fiscal Year 1973,* hearing before the Committee on Foreign Relations, United States Senate, March 20–21 and 28, 1972, Washington, D.C., U.S. Government Printing Office, 1972, p. 55. For a description of the USIA by one of its officials, see A. C. Hansen, *USIA: Public Diplomacy in the Computer Age,* New York, Praeger, 1984. For a critical vision of the Voice of America: See H. Fredericks, *Cuban-American War,* Norwood, N.J., Ablex, 1985. On the U.S. government's propaganda strategies, see Yves Eudes, *La conquête des esprits,* Paris, Maspero, 1982.

29. U.S. Senate, *USIA Appropriations,* session of March 20, 1972, p. 55.

30. Ibid., pp. 7–8.

31. L. Farago, *War of Wits: The Anatomy of Espionage and Intelligence,* Funk & Wagnalls, 1954, p. 330.

32. U.S. Senate, *Foreign and Military Intelligence. Book I, Final Report of the Select Committee to Study Governmental Operations with Respect to Intelligence Activities,* April 26, 1976, Washington, D.C., U.S. Government Printing Office, 1976, pp. 197–98.

33. R. H. Schultz and R. Godson, *Dezinformatsia: Active Measures in Soviet Strategy,* Oxford, Pergamon-Brassey, 1984.

34. U.S. Senate, *Foreign and Military Intelligence,* p. 539.

35. U.S. Senate, *USIA Appropriations,* p. 302.

36. J. Medvedev, *Le secret de la correspondance est garanti par la loi,* Paris, Julliard, 1972.

37. G. Arbatov, *The War of Ideas in Contemporary International Relations: The Imperialist Doctrine, Methods and Organization of Foreign Propaganda,* Moscow, Progress Publishers, 1973 (original edition in Russian, 1970). For an analysis of several Soviet works of this type, see J. R. Bennett, "Soviet Scholars Look at U.S. Media," *Journal of Communication* 36, no. 1, Winter 1986.

38. R. A. Medvedev, "Information in Russia: Against the Languid Flow," *International Herald Tribune,* February 29, 1984, p. 4.

39. J. Der Derian, "Arms, Hostages and the Importance of Shredding in Earnest: Reading the National Security Culture," in *Cultural Politics in Contemporary America,* I. Angus and S. Jhally, coordinators, London and New York, Routledge, 1988.

40. E. E. Dennis et al., *The Media at War: The Press and the Persian Gulf Conflict: A Report of the Gannett Foundation,* New York, Columbia University, 1991.

41. D. C. Hallin, *The "Uncensored War": The Media and Vietnam,* New York, Oxford University Press, 1986.

42. D. C. Hallin, "Living Room War: Then and Now," *Extra!* 4, no. 3 (special issue on the Gulf War), May 1991, p. 21.

43. D. Hertz, "The Radio Siege of Lorient," *Hollywood Quarterly,* no. 1, 1946.

44. E. Canetti, *Masse und Macht,* Hamburg, Claasen Verlag, 1960. English translation: *Crowds and Power,* trans. Carol Stewart, London, Gollancz, and New York, Viking Press, 1962.

45. Voir P. Class, "Sous le feu de la technologie," *01 Informatique,* January 25, 1991; S. Rosselin, "Conflit du Golfe: les satellites savent tout," *Science et Vie,* September 1990.

46. F. Came, "Le casse-tête de la logistique alliée," *Libération,* January 28, 1991.

47. P. Virilio, "Guerre électronique," *Terminal,* no. 11, 1984 (paper read at the symposium "Mythes et imageries de la technologie," June 3–4, 1982).

48. J. Binde, "Le terrorisme technocratique," *Le Monde,* July 13, 1982.

49. F. Rousselot, "L'évènement," *Libération,* March 28, 1991.

6. From Progress to Communication: Conceptual Metamorphoses

1. M. McLuhan, "At the Moment of Sputnik the Planet Became a Global Theater in Which There Are No Spectators but Only Actors," *Journal of Communication* 24, no. 1, Winter 1974, p. 57.

2. M. McLuhan and Q. Fiore, *War and Peace in the Global Village,* New York, Bantam, 1968, pp. 128, 136.

3. A. Uslar Pietri, "Las comunicaciones como revolución," *Vision* 40, no. 8, April 22, 1972.

4. E. B. Weiss, "Advertising Nears a Big Speed-up in Communications Innovation," *Advertising Age,* March 19, 1973, p. 52.

5. See the work of a classical author in this genre, that is, one who has little sense of nuance: Jean-François Revel, *Ni Marx, ni Jésus.* Paris, Laffont, 1971. For a close analysis of anticommunism and its media construction in Europe in the 1970s, one may read the works (of which there is no equivalent in France) of P. Elliott and P. Schlesinger, "Some Aspects of Communism as a Cultural Category," *Media, Culture and Society* 1, 1979, pp. 195–210; D. Childs (ed.), *The Changing Face of Western Communism,* London, Croom Helm, 1980.

6. J. W. Carey, "McLuhan and Mumford: The Roots of Modern Media Analysts," *Journal of Communication* 31, no. 3, Summer 1981.

7. M. McLuhan, *The Mechanical Bride: Folklore of Industrial Man,* New York, Vanguard Press, 1951.

8. L. Mumford, *Technics and Civilization,* New York, Harcourt, Brace & World, 1934.

9. H. Innis, *The Bias of Communication,* Toronto, Toronto University Press, 1951; *Empire and Communications,* Toronto University Press, 1972.

10. M. McLuhan, *The Gutenberg Galaxy: The Making of Typographic Man,* Toronto, University of Toronto Press, 1962, pp. 63–64.

11. T. W. Cooper, "McLuhan and Innis: The Canadian Theme of Boundless Exploration," *Journal of Communication* 31, no. 3, Summer 1981, p. 155.

12. Z. Brzezinski, *Between Two Ages: America's Role in the Technetronic Era,* New York, Viking Press, 1970, p. 13.

13. D. Bell, *The Coming of Post-Industrial Society,* New York, Basic Books, 1973.

14. H. Kahn and A. J. Wiener, *The Year 2000: A Framework for Speculation on the Next Thirty-three Years,* Croton-on-Hudson, N.Y., Hudson Institute, 1967.

15. On this genesis, see M. Marien, "Les deux visions de la société post-industrielle," *Futuribles,* no. 12, Fall 1977.

16. D. Bell, "The Measurement of Knowledge and Technology," in *Indicators of Social Change,* ed. E. Sheldon and W. Moore, New York, 1968. Reprinted in *The Coming of Post-Industrial Society,* Bell himself reports that the original formulation of the concept of postindustrial society was presented by him at a forum on technology and social change in Boston in 1962 in a paper that remained unpublished.

17. D. Bell, *The End of Ideology,* New York, The Free Press, 1962, p. 371.

18. S. M. Lipset, *Political Man,* London, Heinemann, 1960; E. Shils, "The End of Ideology?" *Encounter,* November 1955. By the same author: "Ideology and Civility," in *The Intellectuals and the Powers,* Chicago, University of Chicago Press, 1972.

19. Bell, *End of Ideology,* p. 372.

20. A. Lalande, *Vocabulaire technique et critique de la philosophie,* Paris, PUF, 1956.

21. R. Aron, *L'opium des intellectuels,* Paris, Calmann-Levy, 1955.

22. J. La Palombara, "Decline of Ideology: A Dissent and an Interpretation," *American Political Science Review* 60, no. 1, March 1966, p. 14.

23. R. Barthes, *Mythologies,* sel. and trans. A. Lavers, London, Paladin-Grafton Books, 1973, pp. 154–55. On the effacement of ideologies in pluralist democracy, see Pierre Ansart, *Idéologies, conflits et pouvoir,* Paris, PUF, 1977.

24. J. Gould and W. L. Kolb, *Dictionary of the Social Sciences,* New York, The Free Press, 1964, p. 315.

25. F. Fukuyama, "The End of History?" *The Public Interest,* Summer 1989.

26. Z. Brzezinski, *Between Two Ages,* p. 9.

27. Ibid., p. 18.

28. Ibid., p. 19.

29. C. David, "Toward an International Declaration of Interdependence," *Freedom,* February–March 1945.

30. Z. Brzezinski, *Between Two Ages,* p. 19.

31. Ibid., p. 23.

32. Ibid., p. 33.

33. See P. Breton, *Une histoire de l'informatique,* Paris, La Découverte, 1987.

34. N. Wiener, *Cybernetics: Control and Communication in the Animal and the Machine,* Cambridge, Mass., MIT Press, 1948, p. 11.

35. Ibid., pp. 161–62.

36. F. Machlup, *The Production and Distribution of Knowledge in the United States,* Princeton, Princeton University Press, 1962.

37. A. Downs, *An Economic Theory of Democracy,* New York, Harper, 1957. See also the contributions of K. E. Boulding, "The Economics of Knowledge and the Knowledge of Economics," in *Economics of Information and Knowledge,* ed. D. M. Lamberton, Harmondsworth, Penguin, 1971.

38. M. U. Porat, *The Information Economy: Definition and Measurement,* Washington, D.C., July 1977, 9 volumes.

39. M. U. Porat, "Global Implications of the Information Society," *Journal of Communication* 28, no. 1, Winter 1978.

40. E. Parker, "An Information-Based Hypothesis," *Journal of Communication* 28, no. 1, Winter 1978.

41. A. Giraud et al. (eds.), *Les réseaux pensants: Télécommunications et société,* Paris, Masson, 1978.

42. S. Nora and A. Minc, *L'informatisation de la société,* Paris, La Documentation française, 1978, p. 72. (Translated and published in English under the title of *The Computerization of Society: A Report to the President of France,* Cambridge, Mass., MIT Press, 1980. The reference here is to p. 80 of the translation.)

43. J. P. Chamoux, *Information sans frontières,* Paris, La Documentation française, 1980; A. Madec, *Les flux transfrontières des données,* Paris, La Documentation française, 1980.

In the major industrial countries, the second half of the 1970s was particularly fertile in reports on "informational society." Some representative titles: Consultative Committee on the Implications of Telecommunications for Canadian Sovereignty (Clyne Committee), *Telecommunications and Canada,* Hull, Quebec; Canadian Government Publishing Centre, 1979; Australian Telecommunications Commission, *Telecom 2000: An Exploration of the Long-Term Development of Telecommunications in Australia,* Melbourne, Telecom Australia, 1975. On Japan see: Y. Masuda, *The Information Society as Post-Industrial Society,* Bethesda, Md., World Future Society, 1981; A. S. Edelstein et al., *Information Societies: Comparing the Japanese and American Experience,* Seattle, Wash., International Communications Center, School of Communications, University of Washington, 1978.

44. A. Madec, "Comment définir les règles du jeu," *Le Monde Diplomatique*, December 1980.

45. J.-F. Lyotard, *La condition postmoderne*, Paris, Minuit, 1979, pp. 15–16. (Published in English as *The Postmodern Condition*, trans. G. Bennington and B. Massumi, Minneapolis, University of Minnesota Press, 1984. The reference here is to p. 5.)

46. Ibid., p. 30.

47. F. Guattari, "L'impasse postmoderne," *La Quinzaine Littéraire*, February 1–15, 1986, p. 21.

48. J. Ellul, *La technique ou l'enjeu du siècle*, Paris, Colin, 1954. Published in English as *The Technological Society*, trans. J. Wilkinson, intro. R. K. Merton, New York, Vintage, 1964.

49. F. de Rose, "Les progrès scientifiques et techniques: Les problèmes qu'ils posent à l'Ouest," *Revue de l'OTAN*, October 1978.

50. A. Lloyd, "Sweden Fears Data Processing (DP) Reliance," *Datamation*, June 1980.

51. U.S. Senate, *Range Goals in International Telecommunications and Information: An Outline for United States Policy*, Washington, D.C., U.S. Government Printing Office, 1983.

52. P. Dabezies, "Tentations du tiers monde," *Le Monde*, July 2, 1991.

53. C. Raghavan, "North Blocks High-Tech to South," *Third World Resurgence*, no. 9, May 1991.

54. On the history of computers in Brazil, see A. Mattelart and H. Schmucler, *Communication and Information Technologies*, trans. D. Buxton, Norwood, N.J., Ablex, 1985.

55. Z. Brzezinski, "Washington est le seul super-grand," interview by M. Foucher, *Libération*, December 15, 1990, special edition entitled "La nouvelle planète," p. 16.

56. Brzezinski, *Between Two Ages*, p. 124.

7. The Revolution of Rising Expectations

1. R. McNamara, *The Essence of Security*, New York, Harper & Row, 1968, p. 149. On the genealogy of the concept of development, see S. Latouche, "Un concept à approfondir," *Options*, no. 31, June 1990.

2. D. Lerner, *The Passing of Traditional Society: Modernizing the Middle East*, New York, The Free Press, 1958.

3. Some excerpts from the archives of this research were published under the title "Survey of Communications Patterns in Jordan," in *A Psychological Warfare Casebook*, ed. W. E. Daugherty and M. Janowitz, Baltimore, Johns Hopkins University Press, 1958.

4. See R. Samarajiwa, *The Tainted Origins of the Communication and the Development Field: Voice of America and the Passing of Traditional Society*, Department of Communication, Simon Fraser University, Burnaby, B.C., Canada, 1984.

5. D. C. McLelland, *The Achieving Society*, New York, Van Nostrand, 1961; L. Pye, ed., *Communications and Political Development*, Princeton, Princeton University Press, 1963; W. Schramm, *Mass Media and National Development*, Stanford, Calif., Stanford University Press, 1964.

6. B. Hoselitz and W. Moore, eds., *Industrialisation et société*, Paris, UNESCO, 1963.

7. J. Tunstall, *The Media Are American: Anglo-American Media in the World*, London, Constable, 1977, p. 208.

8. A. Inkeles and D. H. Smith, *Becoming Modern*, Cambridge, Mass., Harvard University Press, 1974.

9. M. Weber, *The Protestant Ethic and the Spirit of Capitalism*, trans. T. Parsons, New York, Charles Scribner, 1930.

10. I. de Sola Pool, "Le rôle de la communication dans le processus de la modernisation et du changement technologique," in *Industrialisation et société,* Hoselitz and Moore, p. 287.

11. W. W. Rostow, *The Stages of Economic Growth,* Cambridge, Cambridge University Press, 1960.

12. Lerner, *The Passing of Traditional Society,* p. 47.

13. UNESCO, *Mass Media in the Developing Countries,* Paris, Paper no. 33, 1961.

14. P. Golding, "Media Role in National Development: Critique of a Theoretical Orthodoxy," *Journal of Communication* 24, no. 3, Summer 1974, pp. 45–46.

15. L. W. Pye, *Guerrilla Communism in Malaya: Its Social and Political Meaning,* Princeton, Princeton University Press, 1956.

16. D. Wilson, "Nation-Building and Revolutionary Wars," in *Nation-Building,* ed. K. W. Deutsch and W. J. Foltz, New York, Atherton Press, 1963, p. 84.

17. L. Pye in *The Role of the Military in Underdeveloped Countries,* ed. J. J. Johnson, Princeton, Princeton University Press, 1962, p. 69.

18. Ibid.

19. R. Moore, "Toward a Definition of Military Nation-Building," *Military Review,* July 1973.

20. L. W. Pye, *Politics, Personality, and Nation-Building: Burma's Search for Identity,* New Haven, Conn., Yale University Press, 1962, p. 13 ; see also L. W. Pye and S. Verba, eds., *Political Culture and Political Development,* Princeton, Princeton University Press, 1965.

21. M. Klare, *War without End,* New York, Vintage, 1972, ch. 9 ("The First Line of Defense").

22. H. Mowlana, "Technology Versus Tradition: Communication in the Iranian Revolution," *Journal of Communication* 29, no. 3, Summer 1979.

23. W. Vogt, "We Help Build the Population Bomb," *New York Times,* April 4, 1965, Section 6, p. 120.

24. *Public Papers of Presidents, Lyndon B. Johnson, 1965,* Washington, D.C., Government Printing Office, 1966, vol. 2, p. 705.

25. Population Reference Bureau, *Population Bulletin,* Washington, D.C., vol. 21, May 1965.

26. See B. Berelson, ed., *Family Planning and Population Programs: A Review of World Development,* Chicago, University of Chicago Press, 1965.

27. J. M. Stycos, "Survey Research and Population Control in Latin America," *The Public Opinion Quarterly* 28, Fall 1964, p. 368.

28. See, as an illustration, D. Bogue, "Some Tentative Recommendations for a 'Sociologically Correct' Family Planning Communication and Motivation Program," in *Research in Family Planning,* ed. C. Kiser, Princeton, Princeton University Press, 1962.

29. For a contradictory analysis of these policies, compare: N. J. Demerath, *Birth Control and Foreign Policy: The Alternatives to Family Planning,* New York, Harper & Row, 1976; and B. Mass, *Population Target: The Political Economy of Population Control in Latin America,* Toronto, Latin American Working Group (LARU), 1976.

30. E. M. Rogers, *Diffusion of Innovation,* New York, The Free Press of Glencoe, 1962; E. M. Rogers and L. Svenning, *Modernization Among Peasants: The Impact of Communication,* New York, Holt, Rinehart & Winston, 1969.

On the history of diffusionist theories in nineteenth-century ethnology, see R. Löwie, *The History of Ethnological Theory,* New York, Holt, Rinehart & Winston, 1937.

31. E. M. Rogers, *Family Planning,* New York, The Free Press, 1973.

32. R. H. Crawford and W. B. Ward, eds., *Communication Strategies for Rural Devel-*

opment, Proceedings of the Cornell-CIAT 1974 International Symposium, Ithaca, N.Y., Cornell University, 1974.

33. L. R. Beltran, "Alien Premises, Objects, Methods in Latin American Communication Research," *Communication Research* 3, no. 2, 1976.

34. P. Freire, *Pedagogy of the Oppressed,* trans. M. B. Ramos, New York, Seabury Press, 1971.

35. "India: Report of United Nations Advisory Mission," *Studies in Family Planning,* June 1966.

36. NASA, *Memorandum of Understanding Between the Department of Atomic Energy of the Government of India and the United States,* 1969.

37. A. Frutkin, "Space Communications in the Developing Countries," in *Communications Technology and Social Policy,* ed. G. Gerbner et al., New York, John Wiley, 1973.

38. N. D. Jayaweera, *Communication Satellites: A Third World Perspective,* Report to the Seminar on New Technologies and the New World Information Order, Bonn-Bad-Godesberg, 1982. Also, see the articles on the SITE experiment in the *Journal of Communication* 29, no. 4, Fall 1979.

39. V. Singh, "Arvin Shinde, le maharajah de la vidéo," *Libération,* April 20, 1987.

40. J. Srampickal, "Are TV and VCRs Threatening Radio in India?," *Media Development,* no. 4, 1990.

41. E. McAnany and J. B. Oliveira, *The SACI/EXERN Project in Brazil: An Analytical Case Study,* Paris, UNESCO, 1980.

42. L. Garcia Dos Santos, *Les dérèglements de la rationalité. Etude sur la démarche systémique du projet SACI/EXERN,* Paris, Université de Paris VII, 1980 (doctoral dissertation in information and communication sciences under the supervision of A. Mattelart).

43. Ibid.

44. See M. and A. Mattelart, *The Carnival of Images,* New York, Bergin & Garvey (Greenwood Press), 1990.

45. J. Baudrillard, "La morale des objets," *Communications,* no. 13, 1969, p. 31.

46. E. M. Rogers, "Communication and Development: The Passing of the Dominant Paradigm," *Communication Research* 2, no. 2, 1976. By the same author: "The Rise and Fall of the Dominant Paradigm," *Journal of Communication* 28, no. 1, 1978.

47. As an example, see J. Galtung et al., *Self-Reliance: A Strategy for Development,* London, Bogle-L'Ouverture Publications, 1980.

48. J. L. Reiffers et al., *Sociétés transnationales et développement endogène: Effets sur la culture, la communication, l'éducation, la science et la technologie,* Paris, UNESCO, 1981.

49. See, for example, C. Hamelink, *Cultural Autonomy in Global Communication,* New York, Longman, 1983.

8. The International Regulation of Information Flows: Two Colliding Views

1. "Principes régissant l'utilisation par les Etats de satellites artificiels de la terre aux fins de télévision directe internationale," reproduced in *Les nouvelles chaînes,* Paris and Geneva, PUF-Cahiers de l'IUED, 1983.

2. U.S. Senate, *USIA Appropriations Authorization, Fiscal Year 1973,* Washington, D.C., U.S. Government Printing Office, 1972, p. 56.

3. R. Bahro, *L'alternative,* Paris, Stock, 1979, p. 222.

4. Quoted in B. Paulu, *Radio and Television Broadcasting in Eastern Europe,* Minneapolis, University of Minnesota Press, p. 99.

5. On the collision between the culture of entertainment and the culture of "television

shortage" under real socialism, see T. Mattelart, *Télévisions occidentales vers l'Est: Une nouvelle frontière?*, paper presented for DEA diploma at Université de Grenoble 3, October 1990. See also C. Feigelson, *La télévision en Union Soviétique,* Paris, INA/Champ Vallon, 1990.

6. On the genesis of the doctrine of the free flow of information, see H. Schiller, *Communication and Cultural Domination,* White Plains, N.Y., M. E. Sharpe, 1976, chap. 2.

7. K. Marx and F. Engels, *Manifesto of the Communist Party* (1848), Moscow, Progress Publishers, 1975, pp. 46–47.

8. S. Le Guevel, *L'information satellitaire: Logiques et enjeux de l'observation de la terre depuis les satellites,* Rennes, Laboratoire CIDOUEST-Université Rennes 2, Department of Information and Communication, 1987.

9. D. W. Smythe, *Dependency Road: Communications, Capitalism, Consciousness,* Norwood, N.J., Ablex, 1981.

10. M. Chemillier-Gendreau, "Le droit de la mer: Mythes et réalités," *Hérodote,* issue on the geopolitics of the sea, no. 32, 1984.

11. E. Díaz Rangel, *Pueblos sub-informados,* Caracas, Monte Avila Editores, 1976 (original edition 1966). See also H. Mujica, *El imperio de las noticias,* Caracas, Ediciones de la Universidad Central de Venezuela (UCV), 1967.

12. A. Pasquali, *Comunicación y cultura de masas,* Caracas, Monte Avila Editores, 1963. See also L. Silva, *La plusvalía ideológica,* Caracas, Ediciones de la Universidad Central de Venezuela, 1970.

13. For Argentina, see above all E. Veron, *Conducta, estructura y comunicación,* Buenos Aires, Ediciones Tiempo Contemporáneo, 1972 (2d ed.). For Chile, see A. and M. Mattelart, and M. Piccini, *Los medios de comunicación en Chile, la ideología de la prensa liberal,* Santiago, special issue of *Cuadernos de la Realidad Nacional* (CEREN), March 1970. For a history of Latin American research in the field, see O. Capriles, "La nouvelle recherche latino-américaine en communication," *Communication 5,* no. 1, Québec, Université Laval, Fall 1982.

14. A. and M. Mattelart, *De l'usage des médias en temps de crise,* Paris, Alain Moreau, 1979. By the same authors, *Rethinking Media Theory: Signposts and New Directions,* trans. J. Cohen and M. Urquidi, Minneapolis, University of Minnesota Press, 1992.

15. T. Guback, *The International Film Industry: Western Europe and America since 1945,* Bloomington, Indiana University Press, 1969, pp. 202–3.

16. H. Schiller, *Mass Communications and American Empire,* Boston, Beacon Press, 1969. See also A. Wells, *Picture Tube Imperialism? The Impact of U.S. Television in Latin America,* Maryknoll, N.Y., Orbis Books, 1972.

17. K. Nordenstreng and T. Varis, *Television Traffic—A One-Way Street?,* Paris, UNESCO, 1974.

18. J. Galtung and M. H. Ruge, "The Structure of Foreign News," *Journal of International Peace Research,* no. 1, 1965. See also J. Galtung, "A Structural Theory of Imperialism," *Journal of Peace Research,* no. 2, 1971.

19. P. Elliott and P. Golding, "The News Media and Foreign Affairs," in *The Management of Britain's External Relations,* ed. R. Boardman and A. J. Groom, London, Macmillan, 1973.

20. J. D. Halloran, P. Elliott, and G. Murdock, *Demonstrations and Communication,* Harmondsworth, Penguin, 1970.

21. G. Cumberbatch and D. Howitt, "Social Communication and War: The Mass Media," in *La communication sociale et la guerre, Etudes de sociologie de la guerre* (colloquium held in May 1974), Brussells, Bruylant, 1974.

22. See, for example, J. Tunstall, *The Media Are American,* London, Constable, 1977;

C. Hamelink, ed., *The Corporate Village: The Role of Transnational Corporations in International Communication,* Rome, IDOC International, 1977; A. Smith, *The Geopolitics of Information,* London, Faber & Faber, 1980; J. O. Boyd-Barrett, *The International News Agencies,* London, Constable, 1980 (based on a dissertation defended in 1976); J. O. Boyd-Barrett and M. Palmer, *Le trafic des nouvelles: Les agences mondiales d'information,* Paris, Alain Moreau, 1981.

23. P. Baran, *The Political Economy of Growth,* Harmondsworth, Penguin, 1957. We refer here to the classic studies by North American authors Paul Sweezy and Immanuel Wallerstein; Brazilian authors Vania Bambirra, F. H. Cardoso, Theotonio Dos Santos, Celso Furtado, and Ruy Mauro Marini; the Chilean Osvaldo Sunkel; the Egyptian Samir Amin; and Europeans André Gunder Frank, Johan Galtung, Pierre Jalée, Christian Palloix, and Dieter Senghaas.

24. I. Wallerstein, *Historical Capitalism,* London, Verso, 1983, p. 30.

25. See A. Mattelart, *Multinational Corporations and the Control of Culture,* Sussex, Harvester Press; Atlantic Highlands, N.J., Humanities Press, 1979 (original French edition 1976); H. Bourges, *Décoloniser l'information,* Paris, Karthala, 1978; Y. Mignot-Lefébvre, ed., Dossier "Audiovisuel et Développement," *Revue Tiers-Monde,* PUF-IEDES, July–September 1979. For a critical evaluation of French research in the 1970s, see P. Flichy, "Current Approaches to Mass Communication Research in France," *Media, Culture and Society,* no. 2, 1980.

26. M. Vovelle in the *Rapport Godelier, Les sciences de l'homme et de la société en France,* Paris, La Documentation française, 1982, p. 258.

27. J. Rigaud, *Les relations culturelles extérieures,* Paris, La Documentation française, 1979, pp. 12, 24. We have quoted this same text in *Rethinking Media Theory* where we analyze in greater detail the different reasons for the accumulated lag in this sector of research.

28. See chapter 6, above.

29. J. O. Boyd-Barrett, "Media Imperialism: Towards an International Framework for an Analysis of Media Systems," in *Mass Communication and Society,* ed. J. Curran et al., London, Arnold, 1977, p. 117.

30. Schiller, *Communication and Cultural Domination,* p. 9.

31. C. C. Lee, *Media Imperialism Reconsidered: The Homogenizing of Television Culture,* Beverly Hills, Calif., Sage, 1980.

32. A. Gramsci, "Analysis of Situations, Relations of Force," *Communication and Class Struggle,* vol. 2, ed. A. Mattelart and S. Siegelaub, New York, International General, 1983, pp. 108–112.

33. A. and M. Mattelart, and X. Delcourt, *International Image Markets: In Search of an Alternative Perspective,* trans. D. Buxton, London, Comedia-Methuen, 1984, pp. 25–26. (Published originally in French, 1983). See also A. Mattelart, "Memorandum for an Analysis of the Cultural Impact of Transnational Firms," in *Transnationals and the Third World: The Struggle for Culture,* trans. D. Buxton, South Hadley, Mass., Bergin & Garvey, 1983, pp. 1–26.

34. See B. Pavlic and C. Hamelink, *Interrelationship between the New International Economic Order and a New International/World Information-Communication Order,* Paris, UNESCO, 1984.

35. See UNESCO, *Communication and Society: A Documentary History of a New World Information and Communication Order 1975–1986,* Paris.

36. See C. Roach, "The U.S. Position on the New World Information and Communication Order," *Journal of Communication* 37, no. 4, 1987; and "The Position of the Reagan Administration on the NWICO," *Media Development* 34, no. 4, 1987.

37. On the Soviet doctrine, see Y. Kaslev, UNESCO and the Soviet Union, Moscow, Novosti, 1986.

38. International Commission for the Study of Communication Problems (MacBride Commission), *Many Voices, One World,* Paris, UNESCO, 1980.

39. For an evaluation: W. E. Preston et al., eds., *Hope and Folly: The United States and UNESCO: 1945–1985,* Minneapolis, University of Minnesota Press, 1990. The *Journal of Communication* has published several issues on the subject, in particular that of Fall 1984 (vol. 34, no. 4). On the recent evolution of the "New Order," see the issue of *Media, Culture and Society* devoted to the question "Farewell to NWICO?," July 1990 (vol. 12, no. 3).

40. U.S. Senate, *Range Goals in International Telecommunications and Information.*

41. O. Capriles, "From National Communication Policies to the New International Information Order: The Role of Research," in *New Structures of International Communication? The Role of Research,* papers from the 1980 Caracas Conference, International Association for Mass Communication Research (IAMCR-AIERI), Leicester, U.K., Adams Bros. & Shardlow, 1982, pp. 36–37.

42. S. Clarke, *Les racines du reggae,* Paris, Editions Caribéennes, 1981, p. 101. (Originally published in English by Heinemann, 1980.)

43. G. Corm, "Au rebours du développement," *Le Monde Diplomatique,* November 1980.

44. UNESCO, World Conference on Education, Jomtien, Thailand, 1990.

45. F. Clairmonte, "Les services, ultimes frontières de l'expansion pour les multinationales," *Le Monde Diplomatique,* January 1991.

46. See chapter 10, below.

47. C. Raghavan, *Recolonisation, l'avenir du tiers monde et les négociations internationales du GATT,* Paris, L'Harmattan-Oxfam, 1990. (Simultaneously published in English by Oxfam, London.)

48. A. Moreau, *Rapport sur l'état de la production audiovisuelle française et européenne au ministre des affaires étrangères,* Paris, 1991.

9. The State in Its Ordinary Dimension

1. H. Cleveland, "La troisième phase de l'Alliance," *Revue de l'OTAN,* no. 6, December 1978.

2. T. W. Adorno and M. Horkheimer, *Dialectic of Enlightenment,* trans. J. Cumming, New York, Herder & Herder, 1972, p. 121.

3. Ibid., p. 154.

4. W. Benjamin, "The Work of Art in the Age of Mechanical Reproduction," *Illuminations,* ed. H. Arendt, New York, Schocken, 1969.

5. A. Mattelart and J. M. Piemme, *Télévision: Enjeux sans frontières,* Grenoble, Presses Universitaires de Grenoble, 1980, p. 15. I have restated here the analyses developed in this earlier study.

6. H. Lefebvre, *De l'Etat,* Paris, Collection 10/18, vol. 4, 1978 ("Les contradictions de l'Etat moderne"), p. 25.

7. Ibid., p. 37.

8. M. Foucault, "La gouvernementalité: Texte d'une leçon," *Actes, Les Cahiers d'Action Juridique,* no. 54, 1986, p. 16. (This text originates from a lecture given in Italy in the late 1970s.)

9. Document from the Conseil Europe, Council of Cultural Cooperation. Report to the preparatory meeting of the conference on the role of the state with respect to cultural industries, Strasbourg, October 9–10, 1978.

10. Conference of European government ministers responsible for cultural affairs, resolutions concerning cultural industries adopted at the Conference of Athens, October 24–26, 1978.

11. A. Girard, "Cultural Industries: A Handicap or a New Opportunity for Cultural Development," in *Cultural Industries: A Challenge for the Future of Culture,* Paris, UNESCO, 1982, p. 27. By the same author: "Industries culturelles," *Futuribles,* no. 17, September 1978.

12. Girard, "Cultural Industries," p. 30.

13. A. Lefèbvre, A. Huet, J. Ion, B. Miège, and R. Péron, *Capitalisme et industries culturelles,* Grenoble, Presses Universitaires de Grenoble, 1978. See also P. Flichy, *Les industries de l'imaginaire,* Grenoble, INA-Presses Universitaires de Grenoble, 1980.

14. On the genesis and evolution of economic analysis of cultural products and services in France, see B. Miège, *The Capitalization of Culture,* New York, International General, 1988.

15. N. Garnham, "Contribution to a Political Economy of Mass Communication," *Media, Culture and Society* 1, no. 2, April 1979.

16. J. Rigaud, *Les relations culturelles extérieures,* Paris, La Documentation française, 1979, p. 66.

17. Ibid., p. 78.

18. Ibid., p. 24.

19. See J. G. Lacroix and B. Lévesque, "Principaux thèmes et courants théoriques dans la littérature scientifique en communication au Québec," *Communication* 7, no. 3, 1985; Dossier, "Les industries culturelles: un enjeu vital!," *Cahiers de recherche sociologique* 4, no. 2, Fall 1986; G. Tremblay, ed., *Les industries de la culture et de la communication au Québec et au Canada,* Montréal, Presses de l'Université du Québec, 1990. Another country where the concept of cultural industry has developed significant roots in the 1980s is Spain. See E. Bustamante and R. Zallo, eds., *Las industrias culturales en España,* Madrid, Akal, 1988.

20. C. Julien, "Les deux bouts de la chaîne...et le milieu," *Alternatives,* issue devoted to the local press, 4th quarter, 1977.

21. I. de Sola Pool, "The Rise of Communications Policy Research," *Journal of Communication* 24, no. 2, Spring 1974, p. 31.

22. H. Mowlana, "Trends in Research on International Communication in the United States," *Gazette* 19, no. 2, 1974. H. Schiller, "Waiting for Orders: Some Current Trends in Mass Communications Research in the United States," *Gazette* 20, no. 1, 1974.

23. Intergovernmental Conference on Communications Policies in Latin America and the Caribbean, Final Report, San José, Costa Rica, July 12–21, UNESCO, Paris, 1976.

24. L. R. Beltran, "Políticas nacionales de comunicación en América latina: los primeros pasos," *Nueva Sociedad* 25, July–August 1976, p. 14. See also M. Tehranian, ed., *Communications Policy for National Development,* London, Routledge & Kegan Paul, 1977.

25. O. Capriles, *El Estado y los medios de comunicación en Venezuela,* Caracas, Ininco-Universidad Central de Venezuela, 1976.

26. O. Capriles, "From National Communication Policies to the New International Information Order: The Role of Research," in *New Structures of International Communication? The Role of Research,* papers from the Caracas Conference, International Association for Mass Communication Research, Leicester, U.K., Adams Bros. & Shardlow, 1982, pp. 34–35.

27. P. F. Lazarsfeld, "Remarks on Administrative and Critical Communication Research," *Philosophy and Social Sciences* 9, 1941.

28. E. Katz, *Social Research on Broadcasting: Proposals for Further Development* (report to the British Broadcasting Corporation), London, BBC, 1977.

29. J. D. Halloran, "Further Development—or Turning the Clock Back?," *Journal of Communication* 28, no. 2, Spring 1978.

30. J. W. Carey, "The Ambiguities of Policy Research," *Journal of Communication* 28, no. 2, Spring 1978.

31. Ibid.

32. French Ministry of Research and Technology, excerpts from a report on the mission to promote the electronics sector, mission presided over by A. Farnoux, Paris, March 1982.

33. J. Barreau et al., *La filière électronique française. Miracle ou mirage?*, Paris, Hatier, 1986.

34. A. Farnoux, *L'électronique dans le monde: Positions 1989–prospectives 1995*, Paris, Electronics International Corporation, November 1990.

35. J. Barreau and A. Mouline, *L'industrie électronique française: 29 ans de relations Etat-groupes industriels (1958–1986)*, Paris, LGDJ, 1987. For a comparative study of French and British communication and industrial policies, see J. Tunstall and M. Palmer, *Liberating Communication: Policy-Making in France and Britain*, London, Basil Blackwell, 1990.

36. E. Le Boucher, "Le média froid de la modernité," *Le Monde*, January 7, 1986, p. 36. By the same author, in collaboration with J. H. Lorenzi, see *Mémoires volées*, Paris, Ramsay, 1979.

37. A. Touraine, *La société post-industrielle*, Paris, Seuil, 1969.

38. A. Touraine, "D'un coup de pied, le plongeur . . . ," *Le Monde*, December 30, 1986, p. 2.

39. F. Mitterrand, *Technologie, emploi, croissance*, Paris, La Documentation française, 1982.

10. The Ascendancy of Geoeconomy: The Quest for Global Culture

1. See A. Mattelart, *Advertising International: The Privatisation of Public Space,* trans. M. Chanan, London-New York, Routledge, Comedia Series, 1991.

2. T. J. Peters and R. Waterman, *In Search of Excellence,* New York, Warner, 1982. On the publishing boom in the English language, see the articles in *Business Week,* January 20, 1986, under the title of "Business Fads: What's In—and Out."

3. M. Vilette, *L'homme qui croyait au management,* Paris, Seuil, 1988.

4. As an illustration, see F. Carmagnola, "Estetica e organizzazione," *Sviluppo e organizzazione,* Milan, November–December 1989; R. Cooper, "Modernism, Post-Modernism and Organizational Analysis: The Contribution of Jacques Derrida," *Organizational Studies* 10, no. 4, 1989.

5. For a caustic commentary on the uses of new paradigms in economics, see C. Durand, "Paradigmes sans rivages," *Les Papiers 5*, Spring–Summer 1989, Toulouse, Presses Universitaires du Mirail.

6. See Mattelart, *Advertising International.* See also J.-M. Charon, ed., *L'état des médias,* Paris, La Découverte, 1991.

7. T. Levitt, "The Globalization of Markets," *Harvard Business Review,* June 1983.

8. J. Becker, coordinator, *Transborder Data Flow and Development,* Bonn, Friedrich-Ebert-Stiftung, 1987. Also note the contributions by geographers in H. Bakis, coordinator, *Communications et territoires,* Paris, La Documentation française, 1990.

9. See, in particular, G. S. Yip et al., "How to Take Your Company to the Global Market," *Columbia Journal of World Business,* Winter 1988.

10. K. Ohmae, "Planting for a Global Harvest," *Harvard Business Review,* July–August 1989, p. 139. See also, by the same author, "Managing in a Borderless World," *Harvard Business Review,* May–June 1989.

11. Ohmae, "Planting for a Global Harvest."

12. A. Bressand, in *L'Expansion,* November 23–December 6, 1989.

13. C. Lorenz, "La fin du centralisme," excerpt from *Financial Times* in *Courrier International,* no. 7, December 20–26, 1990.

14. G. Beney, "Travail planétaire et chômage humain," *Terminal,* July–August 1989, p. 21.

15. K. Ohmae, *Triad Power,* New York, Free Press, 1985.

16. See A. Mattelart and M. Palmer, "Advertising in Europe: Promises, Pressures and Pitfalls," *Media, Culture & Society* 13, no. 4, October 1991.

17. J. Habermas, *The Structural Transformation of the Public Sphere: An Inquiry into a Category of Bourgeois Society,* Cambridge, Mass., MIT Press, 1989.

18. A. Berque, "La communication intergroupale au Japon," in *Information et communication,* Séminaires interdisciplinaires du Collège de France, ed. A. Lichnerowitz et al., Paris, Maloine, 1983.

19. See M. Raboy and B. Dagenais, eds., *Media, Crisis and Democracy: Mass Communication and the Disruption of Social Order,* London, Sage, 1992.

20. A. Kahn, "Les technologies américaines menacées," *Le Monde,* March 23, 1991.

21. P. Sabatier, "Japon/Etats-Unis," *Libération,* July 6–7, 1991.

22. J. Mousseau, "Plaidoyer pour une industrie française du dessin animé," *Communication et Langages,* no. 52, 1982.

11. Mediations and Hybridizations: The Revenge of the Cultures

1. L. Boltanski, *Les cadres: La formation d'un groupe social,* Paris, Minuit, 1982, p. 155. Translated as: *The Making of a Class: Cadres in French Society,* trans. Arthur Goldhammer, Cambridge, Cambridge University Press, 1987.

2. Philippe d'Iribarne, *La logique d'honneur: Gestion des entreprises et traditions nationales,* Paris, Seuil, 1989, p. 9.

3. Ibid., p. 263.

4. For example: J. Ruffier, "La gestion de l'automatisation: Un modèle mexicain," *Revue Française de Gestion,* no. 64, September–October 1987; Dossier, "Marketing et pays en voie de développement," *Revue Française de Marketing,* no. 2, 1987.

5. P. Fridenson and A. Straus, eds., *Le capitalisme français,* Paris, Fayard, 1987.

6. N. Alter, "La participation: Piège à innovation?," Prospective et Télécom, *Lettre du service de la prospective et des études économiques (SPES),* Paris, *Direction générale des télécommunications,* no. 11, May 1987.

7. Alter, "La participation."

8. F. Guattari, "Les nouveaux mondes du capitalisme," *Libération,* December 22, 1987.

9. M. Crozier et al., Introduction to *The Crisis of Democracy,* New York, New York University Press, 1975, pp. 7–8.

10. M. Crozier et al., "Appendix I. Discussion of Study during Plenary Meeting of the Trilateral Commission," ibid., p. 182.

11. "Conclusion," ibid., p. 161.

12. D. Bell, "Notes on the Post-Industrial Society," *The Public Interest,* Winter 1967.

13. On the evolution of the intellectual classes in the major industrial countries, see R. Miliband and L. Panitch, coordinators, *The Retreat of the Intellectuals,* London, Merlin

Press, 1990; A. and M. Mattelart, Part 3: "Intellectuals and Media Culture: Redefining the Relationship," in *Rethinking Media Theory*, trans. J. Cohen and M. Urquidi, Minneapolis, University of Minnesota Press, 1992; G. Vattimo, *La société transparente*, Paris, Desclée de Brouwer, 1990.

14. G. Deleuze, "Les sociétés de contrôle," *L'Autre Journal*, May 1990. By the same author, *Pourparlers*, Paris, Minuit, 1990.

15. U. Hannerz, "Notes on the Global Ecumene," *Public Culture* 1, no. 2, Spring 1989.

16. A. Appadurai, "Disjuncture and Difference in the Global Cultural Economy," *Public Culture* 2, no. 2, Spring 1990, p. 16.

17. R. Ortiz, *A moderna tradiçao brasileira*, São Paulo, Editora Brasiliense, 1988.

18. M. and A. Mattelart, *The Carnival of Images: Brazilian Television Fiction*, New York, Bergin & Garvey (Greenwood Press), 1990.

19. Ortiz, *A moderna tradiçao brasileira*, p. 206.

20. M. Castells, "Le commencement de l'histoire," *Le Socialisme du Futur* 1, no. 2, 1990. By the same author, "High Technology and the New International Division of Labour," *International Labour Review*, October 1989; and, in collaboration with R. Laserna, "The New Dependency," *Sociological Forum*, Summer 1990.

21. See A. Mattelart, *Advertising International: The Privatisation of Public Space*, trans. M. Chanan, London-New York, Routledge, Comedia Series, 1991, chaps. 8 and 9.

22. M. Foucher, in *Libération*, special edition entitled "La Nouvelle Planète."

23. H. Salazar del Alcazar, "El teatro peruano de los 80: las marcas de la historia y de la violencia de estos días," *Conjunto/Casa de las Américas*, January–March 1990, p. 38.

24. P.-A. Taguieff, *La force du préjugé*, Paris, La Découverte, 1988.

25. By way of illustration: J. Martín Barbero, *De los medios a las mediaciones*, Barcelona, Gustavo Gili, 1987; A. Rajadhyaksha, "Neo-Traditionalism: Film as Popular Art in India," *Framework*, no. 32, 1986. The journal *Telos* (Madrid) published a monographic issue on new trends in research in Latin America (September–November 1989), as did *Media, Culture & Society* (no. 4, 1988).

26. See R. Kothari, "New Social Forces," *Development Seeds of Change*, no. 1, 1985; P. Waterman, "Is the People's Flag Deepest Red . . . or Brightest Green? Reflections on the New Social Movements Internationally," *Philippine Quarterly of Third World Studies* 2, no. 3, 1987. Also see the final declaration of the International Conference of International Organizations: "The Manila Declaration on People's Participation and Sustainable Development," *IFDA Dossier*, Nyon, Switzerland, January–April 1990. Among the numerous publications illustrating the link between network and communication, see M. Gutierrez, coordinator, *Video, tecnología y comunicación popular*, Lima, IPAL-Crocevía, 1989.

27. W. Schramm, "The Unique Perspective of Communication: A Retrospective View," *Journal of Communication* 33, no. 3, Summer 1983, p. 15.

28. E. Katz and T. Liebes, "Mutual Aid in the Decoding of *Dallas*," in P. Drummond and R. Paterson, coordinators, *Television in Transition*, London, British Film Institute (BFI), 1985.

29. See in particular E. Katz, M. Gurevitch, and H. Haas, "On the Uses of the Mass Media for Important Things," *American Sociological Review* 38, 1973; J. Blumler and E. Katz, coordinators, *The Uses and Gratification Approach to Mass Communication Research*, Sage Annual Review of Communication Research, vol. 3, Beverly Hills, Calif., Sage, 1975.

30. J. M. Piemme, *La télévision comme on la parle*, Brussels-Paris, Editions Labor/Fernand Nathan, 1978, p. 95.

31. E. Katz, "A propos des médias et de leurs effets," in *Technologies et symboliques de la communication*, ed. L. Sfez et al., Grenoble, Presses Universitaires de Grenoble, 1990.

32. To fuel the debate on the reconceptualization of the audience, see J. Fiske, *Television Culture,* London-New York, Methuen, 1987; J. Curran, "The New Revisionism in Mass Communication Research: A Reappraisal," *European Journal of Communication* 5, no. 2–3, June 1990; D. Morley, *Television Audiences: Eight Cultural Studies,* London, Routledge, 1992.

33. For a critique, see P. Dahlgren, "Media, Meaning and Method: A 'Post-Rational' Perspective," *Nordicom Review,* no. 2, 1985; I. Bondebjerg, "Critical Theory, Aesthetics and Reception Research," *Nordicom Review,* no. 1, 1988.

34. R. Hoggart, *The Uses of Literacy: Changing Patterns in English Mass Culture,* Fairlawn, N.J., Essential Books, 1957.

35. On the genealogy of research in Great Britain, see N. Garnham, "Toward a Theory of Cultural Materialism," *Journal of Communication* 33, no. 3, Summer 1983. On approaches to reception through literary studies see, for example, T. Todorov, *Mikhail Bakhtine: Le principe dialogique,* Paris, Seuil, 1980; H. R. Jauss, *Pour une esthétique de la réception,* Paris, Gallimard, 1978.

36. M. de Certeau, *L'invention du quotidien,* vol. 1: *Arts de faire,* Paris, Folio/Essais, 1990 (first edition, 1980). Translated as *The Practice of Everyday Life,* Berkeley, University of California Press, 1987. On networks, see M. de Certeau and L. Giard, *L'ordinaire de la communication,* Paris, Dalloz, 1983.

37. Quoted in S. Gaubert, *Proust ou le roman de la différence,* Lyon, Presses Universitaires de Lyon, 1980.

Conclusion: The Enigma

1. Declaration by intellectuals of Spain, the United States, and Latin America assembled in Mexico City, January 6–7, 1991.

2. E. Canetti, *Crowds and Power,* trans. Carol Stewart, New York, Viking Press, 1962, p. 169.

3. J. L. Borges, "The Congress," *The Book of Sand,* trans. N. Thomas di Giovanni, London, Penguin, pp. 21–22.

Index

Compiled by Robin Jackson

Armand Mattelart was born in Belgium in 1936. He has authored or coauthored several books on culture, politics, the mass media, and communications theory. Since 1983 he has been full professor of information and communication sciences at the Université de Haute-Bretagne (Rennes-2), France. From 1962 to 1973 he was a professor of sociology of communication at the Catholic University of Chile. He has carried out numerous research missions in Europe, Latin America, and Africa for UNESCO, the French Foreign Ministry, and the Centre National de Recherche Scientifique (CNRS), France. His most recent works include *L'Internationale publicitaire* (1989) and, in collaboration with Michèle Mattelart, *Le carnaval de images. La fiction brésilienne* (1987) and *Penser les medias* (1986), published in English translation as *Rethinking Media Theory: Signposts and New Directions* (Minnesota, 1992).

Susan Emanuel received a doctorate in communications from the University of Rennes, France, in 1992. She has worked as a producer of educational programs for the BBC, and as a lecturer in film and television at the University of Bristol, England. As Susan Boyd-Bowman, she has published a number of articles on American and European film and television, and in the late 1980s was on the editorial board of *Screen*. She lives in Boston and in Brittany, and teaches media studies.

James A. Cohen teaches political science at the University of Lausanne, Switzerland. Born in the United States in 1954, he has lived in France since 1977. He has written extensively on the U.S.-Puerto Rico relationship and on political regimes in Latin America and published articles in *Les Temps Modernes, L'Evénement Européen, Futur Antérieur,* and *Annales des Pays d'Amérique Centrale et de la Caraïbe.* He is cotranslator of an earlier book by Armand and Michèle Mattelart, *Rethinking Media Theory: Signposts and New Directions* (Minnesota, 1992).